FOOD OILS and FATS

FOOD OILS
and
FATS

Technology, Utilization, and Nutrition

HARRY LAWSON

CHAPMAN & HALL

I(T)P An International Thomson Publishing Company

New York • Albany • Bonn • Boston • Cincinnati
• Detroit • London • Madrid • Melbourne • Mexico City
• Pacific Grove • Paris • San Francisco • Singapore
• Tokyo • Toronto • Washington

Cover Design: Andrea Meyer, emDASH Inc.

Copyright © 1995
By Chapman & Hall
A division of International Thomson Publishing Inc.
I(T)P The ITP logo is a trademark under license

Printed in the United States of America

For more information, contact:

Chapman & Hall
One Penn Plaza
New York, NY 10119

International Thomson Publishing
Berkshire House 168-173
High Holborn
London WC1V 7AA
England

Thomas Nelson Australia
102 Dodds Street
South Merlbourne, 3205
Victoria, Australia

Nelson Canada
1120 Birchmount Road
Scarborough, Ontario
Canada, M1K 5G4

Chapman & Hall
2-6 Boundary Row
London SE1 8HN

International Thomson Editores
Campos Eliseos 385, Piso 7
Col. Polanco
11560 Mexico D.F. Mexico

International Thomson Publishing Gmbh
Königwinterer Strasse 418
53228 Bonn
Germany

International Thomson Publishing Asia
221 Henderson Road
#05-10 Henderson Building
Singapore 0315

International Thomson Publishing-Japan
Hirakawacho-cho Kyowa Building, 3F
1-2-1 Hirakawacho-cho
Chiyoda-ku, 102 Tokyo
Japan

1 2 3 4 5 6 7 8 9 10 XXX 01 00 99 97 96 95

Library of Congress Cataloging-in-Publication Data

Lawson, Harry.
    Food oils & fats : technology, utilization, and nutrition / Harry Lawson.
        p.   cm.
    Includes bibliographical references and index.
    ISBN 0-412-98841-0
    1. Oils and fats, Edible. I. Title. II. Title: Food oils and fats.
    TP670.L37   1994
    664'.3—dc20   94-17775
                                             CIP

The use in this book of trademarks, trade names, general descriptive names, and so forth, even if they are not especially identified, shall not be taken as indication that such names, as understood by the Trade Mark and Merchandise Act, may be freely used by anyone.

Please send your order for this or any Chapman & Hall book to **Chapman & Hall, 29 West 35th Street, New York, NY 10001, Attn: Customer Service Department.** You may also call our Order Department at 1-212-244-3336 or fax your purchase order to 1-800-248-4724.

For a complete listing of Chapman & Hall's titles, send your requests to **Chapman & Hall, Dept. BC, One Penn Plaza, New York, NY 10119.**

To all who have an interest in food products whether as a professional in the food field, a student, or as a consumer. Knowledge of Oils and Fats is basic to an understanding of food products, including the selection of the raw materials, proper preparation and processing, and nutritional considerations.

Contents

Preface

This is a basic reference/textbook for professionals and students involved with these important oils and fats. It is a valuable source of information for those preparing for or already professionally associated with the Food Processing and Foodservice industries.

Chapters one through six deal with the technology of oils and fats, including sources, chemical structure, physical and chemical properties, and processing techniques. Chapters seven through twelve are devoted to the utilization of oils and fats in Food Manufacturing and Foodservice, including deep frying, griddling, baking of all types, salad dressings, margarines, hard butters, and dairy product replacements. The last four chapters contain a most complete and up-to-date treatment of nutrition, as well as the latest developments in analytical methods, flavor, and product development as they relate to oils and fats.

This book contains the necessary information for an understanding of how oils and fats are used in the food industry and how this information is used to set standards and meet performance goals. In a thoroughly readable way it is a how-to-do, hands-on treatise on using oils and fats for every major food use.

Acknowledgments

I gratefully acknowledge many friends at Procter & Gamble who provided updated material, some currently employed and some recently retired. Fred J. Baur, formerly of Procter & Gamble, wrote the updated chapters related to Analytical Methods, Flavor, Nutrition, and Dietary Considerations.

The Institute of Shortening and Edible Oils was very generous in permitting us the use of material published in their latest edition of *Food Fats and Oils*.

Marge Bomkamp did her usual most excellent work in processing the manuscript. My devoted wife Betty provided some artwork, proofed copy, monitored grammatical problems, and especially provided moral support.

Chapter 1

Definitions and Overview

This book is a study of oils and fats, with special emphasis on their uses in the food processing and foodservice industries. In order to use these products efficiently, it is necessary to have a good understanding of the properties and behaviors of oils and fats.

Oils and fats constitute one of the three major classes of food products. The other two are proteins and carbohydrates. Oils and fats are essential nutrients of the human diet and are concentrated sources of energy. They supply about 9 kcal/g, as compared with about 4 kcal/g from protein and carbohydrates.

In any study of human nutritional requirements, the contributions of each of these major food classes must be understood in order to set up proper diets for individuals. Knowledge of vitamins and minerals as well as the roles of oxygen, water, fiber, and trace elements is also essential.

Oils and fats occur naturally in many of our foods, such as dairy products, meats, poultry, nuts, fish, and vegetable oil seeds. They are very important in the processing of many favorite foods. Examples include french fried potatoes, salad dressings, an infinite variety of breads and rolls, and dessert items.

Processed foods such as dairy products, margarine, snacks, prefried poultry, peanut butter, coffee whiteners, crackers, and so on contain significant quantities of oils and fats.

All oils and fats are predominantly tri-fatty acid esters of glycerol, commonly referred to as triglycerides. The term "lipids" is all-in-

1

clusive and includes triglycerides, sterols (including cholesterol), phosphatides, monoglycerides, diglycerides, free fatty acids, fatty alcohols, waxes, terpenes, fat-soluble vitamins, and others.

This book is confined to the study of food oils and fats. It does not include inedible products such as waxes, lubricating oils, drying oils for paints, and fats used in soap making.

Oils and fats are basically insoluble in water but soluble in most organic solvents, such as hexane, carbon tetrachloride, petroleum ether, and ethyl ether.

The term "fat" generally refers to substances that are solid or semisolid at normal room temperature (70–75°F, 21.1–23.9°C). "Oils" are liquid at and below these temperatures. However, these terms may be used interchangeably. Examples of fat products include vegetable shortenings, butter, margarine, and lard. Examples of oils are canola, corn, cottonseed, soybean, and sunflower.

Food fats can also be classified by their origin, either plant or animal. Typical examples of plant derivatives include soybean, canola, cottonseed, sunflower, and safflower oils, whereas fats of animal origin include lard and butterfat.

Currently fats and oils provide about 38% of the caloric needs of people in the United States (1). This is another important reason to have a good understanding of these products.

SUMMARY

Studies in this book will be limited to edible food oils and fats, as distinguished from the inedible types. These oils and fats are very important from a nutritional standpoint because as a class they supply an average of 38% of the calories in the diet of the people in the United States.

References

1. Institute Shortening Edible Oils. 1988. *Food Fats and Oils*, 6th ed. Washington, DC: Institute of Shortening and Edible Oils, p. 6.

Chapter 2

The Basic Chemistry of Oils and Fats

The carbon atom is the basic element in food chemistry, including oils and fats. Carbon atoms, with a valence of 4, may bond together with other carbon atoms to form molecules with long chains. Furthermore, carbon's ability to form bonds or react with other elements such as hydrogen, oxygen, iodine, nitrogen, and phosphorus is fundamental to understanding the chemistry of oils and fats.

Basically, oils and fats are mixtures of triglycerides. This is the manner in which they are composed naturally.

$$
\begin{array}{c}
\text{H} \\
| \\
\text{H—C—OH} \\
| \\
\text{H—C—OH} \\
| \\
\text{H—C—OH} \\
| \\
\text{H}
\end{array}
$$

Glycerol

The glycerol molecule has three carbon atoms, together with five hydrogen atoms and three OH or hydroxyl groups. It should be noted that there are four bonds or linkages to each of the three carbon atoms. When three fatty acids are combined with one glycerol molecule, we have a triglyceride.

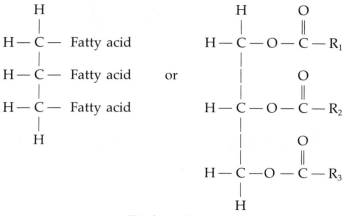

Triglyceride

"R" is the method of abbreviating portions of long-chain radicals such as those of fatty acids. When all of the fatty acids in a triglyceride are identical, it is called a simple triglyceride. However, the much more common forms are the mixed triglycerides in which two or three different fatty acids are present in the molecules.

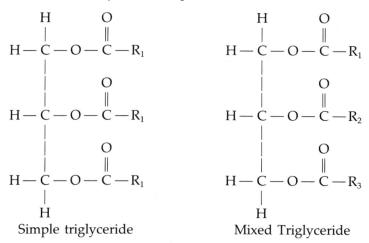

Simple triglyceride   Mixed Triglyceride

If, in the above illustrations, we designate R_1 to be stearic acid, R_2 palmitic acid, and R_3 oleic acid; then the triglyceride on the left would be simply tristearin. The mixed triglyceride on the right would be stearopalmitolein.

If only two fatty acids are attached to a specific glycerol molecule, we have a diglyceride; if only one fatty acid is attached, the molecule is a monoglyceride. Each carbon atom continues to have four linkages.

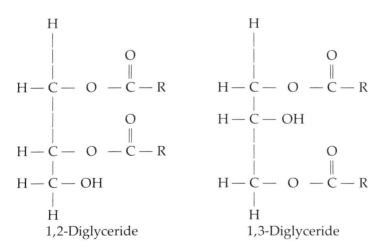

Diglycerides can be further described as a 1,2-diglyceride or a 1,3-diglyceride, depending on the position of fatty acids on the glycerol molecule.

Monoglycerides may be either 1 (or alpha)-monoglyceride or 2 (or beta)-monoglyceride.

```
        H                              H
        |                              |
        |              O        H — C — OH
        |              ||             |
  H — C — O   — C — R               O
        |                            ||
  H — C — OH         H — C — O   — C — R
        |                            |
  H — C — OH         H — C — OH
        |                            |
        H                            H
```

1 (or alpha)-monoglyceride   2 (or beta)-monoglyceride

Mono and diglycerides are important as emulsifiers in food products. Their preparation and use will be discussed more fully later in this book. Monoglycerides and diglycerides are also formed in the intestinal tract as the result of the normal digestion of triglycerides. In addition, they occur naturally in minor amounts in both vegetable oils and animal fats.

Any fatty acid not linked to a glycerol or some other molecule in an oil or fat is referred to as a "free fatty acid." The major component of all fats and oils is triglycerides, representing well over 95% of the weight of most food fats in the form in which they are consumed (1).

Most unrefined oils contain relatively high levels of free fatty acids. A typical level for crude soybean oil is from 0.5 to 1.5%. Crude palm oil may contain 3.0–5.0% free fatty acids. Refined oils and fats that are ready for use in foods usually have a free fatty acid level of less than 0.05%.

Some of the more common fatty acids found in naturally occurring oils and fats are butyric, lauric, palmitic, stearic, oleic, and linoleic. A 1-lb can of shortening, for example, contains innumerable fat molecules consisting of mixtures of the various fatty acids attached to the glycerol molecules. The relative number of these various fatty acids and their particular placement on the glycerol molecules determine the various characteristics of the oil or fat products. The processing techniques employed also affect a product's physical and performance characteristics. All oils and fats are built from a relatively small number of fatty acids.

Some fats are solid at room temperature, whereas others are liquid. Those that are liquid at room temperature are referred to as oils/liquid oils/fluid fats/liquid shortenings. It should be kept in mind that fluid fats/liquid shortenings are included in this grouping even though they contain a small amount of solids at room temperature (usually less than 10%). These liquids also may be referred to as "unsaturated" fats. This does not mean that all of the fatty acids in that particular product are unsaturated, but merely that there is generally a high proportion of unsaturated fatty acids, sufficient to render this specific product liquid. For example, soybean oil has a preponderance of unsaturated fatty acids, making it a liquid, whereas lard has a greater proportion of saturated fatty acids, making lard a solid at room temperature. Coconut oil is a somewhat special case in that it is liquid to just about 78°F (26°C) in spite of the fact that it contains 85–95% saturated fatty acids.

Chain length, or the number of carbon atoms in a fatty acid, also has a great influence on whether a fat is solid or liquid. Most fatty acids have from 4 to 22 carbon atoms, primarily in even numbers. The products containing high proportions of the longer-chain fatty acids (14–22 carbon atoms) are likely to be solid at room temperature, whereas those containing more of the shorter-chain fatty acids (4–12 carbon atoms) are likely to be liquid. Coconut oil is very high in lauric acid, which contains 12 carbon atoms, and about 60–65% of the fatty acids are of 14 carbon atoms or less. This is the reason for its liquidity at relatively low temperatures. Therefore, the most important factors that render a product solid or liquid are the average fatty acid chain length and the amount of saturated in relation to unsaturated fatty acids.

The processing techniques employed and the ultimate crystal structure of the product also have an effect on the product's final physical form. This will be discussed in a later chapter.

$$\begin{array}{ccc} H & H & O \\ | & | & \| \\ R-C-C-C-OH \\ | & | \\ H & H \end{array}$$

Saturated Fatty Acid

In the illustration of a saturated fatty acid, all of the carbon atoms have four linkages to the other atoms, including the other carbon atoms in the molecule. The "R" refers to the balance of the molecule. If this were an 18-carbon molecule, 15 carbon atoms would be included in "R."

The carboxyl group is characteristic of all fatty acids. It is also the portion of the fatty acid molecule that is attached to glycerol to form the monoglyceride, diglyceride, or triglyceride. This carboxyl radical is quite often written as COOH.

$$\text{COOH} \quad \text{or} \quad -\overset{\overset{\displaystyle O}{\|}}{C}-OH$$

Carboxyl Group

Fatty acids are predominantly saturated and unsaturated straight aliphatic chains with an even number of carbon atoms and a carboxyl group as illustrated.

$$R-\overset{\overset{\displaystyle H}{|}}{\underset{\underset{\displaystyle H}{|}}{C}}-\overset{\overset{\displaystyle H}{|}}{\underset{\underset{\displaystyle H}{|}}{C}}-\overset{\overset{\displaystyle O}{\|}}{C}-OH$$

Aliphatic   Carboxyl
Chain   Group

There are chains in which there is a double bond between a pair of carbon atoms in a molecule; then we have an unsaturated fatty acid as shown. This double bond results in a more reactive linkage between these two carbon atoms, which, in turn, results in an ability to add hydrogen (or other elements) at the site of this double bond. In this manner, an unsaturated fatty acid may react with hydrogen to break this double bond and form a saturated fatty acid.

$$R-\overset{\overset{\displaystyle H}{|}}{C}=\overset{\overset{\displaystyle H}{|}}{C}-\overset{\overset{\displaystyle H}{|}}{\underset{\underset{\displaystyle H}{|}}{C}}-COOH$$

Unsaturated Fatty Acid (Monounsaturated or Monoenoic)

A polyunsaturated fatty acid has two or more points of unsaturation in a specific fatty acid molecule. Two double bonds make this molecule more unstable, and it reacts with hydrogen, oxygen, and other elements more readily than a monounsaturated fatty acid.

$$R-\overset{\overset{\displaystyle H}{|}}{C}=\overset{\overset{\displaystyle H}{|}}{C}-\overset{\overset{\displaystyle H}{|}}{\underset{\underset{\displaystyle H}{|}}{C}}-\overset{\overset{\displaystyle H}{|}}{C}=\overset{\overset{\displaystyle H}{|}}{C}-\overset{\overset{\displaystyle H}{|}}{\underset{\underset{\displaystyle H}{|}}{C}}-COOH$$

Polyunsaturated Fatty Acid (or Polyenoic)

Table 2.1. Important Fatty Acids

	Carbon Atoms	Double Bonds	Melting Point °F	Melting Point °C	Major Occurrence in Natural Oils and Fats
Butyric	4	0	18	−8	Butter
Lauric	12	0	111	44	Coconut oil
Myristic	14	0	129	54	Butter, coconut oil, palm oil
Palmitic	16	0	145	63	Palm oil, butter, and meat fats such as chicken fat, lard, tallow
Stearic	18	0	157	69	Tallow, cocoa butter, lard, butter
Oleic	18	1	58	14	Olive, peanut, lard, palm, tallow, corn, rapeseed, canola
Linoleic	18	2	23	−5	Soybean, safflower, sunflower, corn, cottonseed
Linolenic	18	3	12	−11	Soybean, canola
Gadoleic	20	1	—	—	Some fish oils
Arachidonic	20	4	−40	−40	Lard, tallow
	20	5	—	—	Some fish oils
Behenic	22	0	176	80	Peanut, rapeseed
Erucic	22	1	91	33	High erucic acid rapeseed
	22	6	—	—	Some fish oils

A fatty acid with three points of unsaturation in its molecule reacts very rapidly with hydrogen or oxygen.

Table 2.1 lists the names of the most important fatty acids occurring in nature, along with their chain lengths and numbers of double bonds (3). This table shows the fatty acids' common names, which are sufficient for a general understanding of food oils and fats. However, for a more complete understanding of fat structure, the Geneva system of nomenclature shown in Table 2.2 should also be used, as it provides a systematic means of naming these acids (3).

In the Geneva system, a Greek prefix is used to designate the number of carbon atoms. For example, stearic acid, with 18 carbon atoms, would have the prefix "octadec." Furthermore, for saturated fatty acids, one uses the suffix "anoic." Therefore, the Geneva system name for stearic acid becomes octadecanoic acid. For unsaturated fatty acids, the suffix is modified according to the number of

Table 2.2.   Saturated Acids

Common Name	Geneva Name	No. of Carbon atoms	Formula
Acetic	Ethanoic	2	CH_3COOH
Butyric	Butanoic	4	C_3H_7COOH
Caproic	Hexanoic	6	$C_5H_{11}COOH$
Caprylic	Octanoic	8	$C_7H_{15}COOH$
Capric	Decanoic	10	$C_9H_{19}COOH$
Lauric	Dodecanoic	12	$C_{11}H_{23}COOH$
Myristic	Tetradecanoic	14	$C_{13}H_{27}COOH$
Palmitic	Hexadecanoic	16	$C_{15}H_{31}COOH$
Stearic	Octadecanoic	18	$C_{17}H_{35}COOH$
Arachidic	Eicosanoic	20	$C_{19}H_{39}COOH$
Behenic	Docosanoic	22	$C_{21}H_{43}COOH$

points of unsaturation. One double bond is described by the suffix "enoic," two double bonds by "dienoic," and three double bonds by "trienoic." Therefore, linolenic acid with 18 carbon atoms and 3 double bonds becomes octadecatrienoic acid. More complete lists of fatty acids with their Geneva names and formulas are shown in Tables 2.2 and 2.3 (3).

Also in the Geneva system of nomenclature, the carbons in a fatty acid chain are numbered consecutively from the end of the chain, the carbon of the carboxyl group being considered as number 1. By convention, a specific double bond in a chain is identified by the lower number of the two carbons that it joins. In oleic acid (octadecenoic acid), for example, the double bond is between the ninth

Table 2.3.   Unsaturated Acids

Common Name	Geneva Name	No. of Double Bonds	No. of Carbon Atoms	Formula
Myristoleic	Tetradecenoic	1	14	$C_{13}H_{25}COOH$
Palmitoleic	Hexadecenoic	1	16	$C_{15}H_{29}COOH$
Oleic	Octadecenoic	1	18	$C_{17}H_{33}COOH$
Linoleic	Octadecadienoic	2	18	$C_{17}H_{31}COOH$
Linolenic	Octadecatrienoic	3	18	$C_{17}H_{29}COOH$
Arachidonic	Eicosatetraenoic	4	20	$C_{19}H_{31}COOH$
Erucic	Docosenoic	1	22	$C_{21}H_{41}COOH$

and tenth carbon atoms. For linoleic acid (9,12-octadecadienoic acid), the double bonds are between the 9 and 10 carbon atoms and the 12 and 13 carbon atoms.

Another system of nomenclature in use for unsaturated fatty acids is the "omega" or "n minus" classification. "Omega" or "n minus" notation often is used by biochemists to designate sites of enzyme reactivity or specificity. The terms "omega" or "n minus" refer to the position of the double bond of the fatty acid closest to the methyl end of the molecule. The methyl end is the furthest away from the carboxyl end of the fatty acid molecule. Thus, oleic acid, which has its double bond 9 carbons from the methyl end, is considered an omega-9 (or an n-9) fatty acid. Similarly, linoleic acid, common in vegetable oils, is an omega-6 (n-6) fatty acid because its second double bond is 6 carbons from the methyl end of the molecule (i.e., between carbons 12 and 13 from the carboxyl end). Eicosapentaenoic acid, found in many fish oils, is an omega-3 (n-3) fatty acid. Alpha-linolenic acid, found in certain vegetable oils, is also an omega-3 (n-3) fatty acid.

When two fatty acids are identical except for the location of the double bond, they are referred to as positional isomers. Fatty acid isomers are discussed at greater length in Chapter 3.

Because of the presence of double bonds, unsaturated fatty acids are more reactive chemically than are saturated fatty acids. This reactivity increases as the number of double bonds increases.

Although double bonds normally occur in a nonconjugated position, they can occur in a conjugated position (alternating with a single bond) as illustrated below:

```
    H   H   H   H           H   H   H   H   H
    |   |   |   |           |   |   |   |   |
  —C=C—C=C—               —C=C—C—C=C—
                                    |
                                    H
      Conjugated            Nonconjugated
```

With the bonds in a conjugated position, there is a further increase in certain types of chemical reactivity. For example, oils and fats are much more subject to oxidation and polymerization when the double bonds are in the conjugated position.

Tables 2.4 and 2.5 show the typical fatty acid composition of some of the more important, commercially used food oils and fats (4). These composition figures should be considered as typical but not exact. In the case of vegetable oils, the actual values vary, depend-

Table 2.4. **Fatty Acid Composition of Some Vegetable Oils**

		% Soybean	% Cottonseed	% Palm	% Coconut	% Canola
C_8	Caprylic	—	—	—	8	—
C_{10}	Capric	—	—	—	6	—
C_{12}	Lauric	—	—	Trace	49	—
C_{14}	Myristic	Trace	1	1	18	—
C_{16}	Palmitic	12	24	48	8	4
C_{18}	Stearic	4	2	5	3	2
$C_{18:1}$	Oleic	23	18	38	7	62
$C_{18:2}$	Linoleic	53	53	9	2	22
$C_{18:3}$	Linolenic	8	Trace	Trace	—	10

ing on such factors as location of growth area, soil conditions, climate, and growing conditions. Meat fats also vary considerably in accordance with the type of animal, the type of feed, the activity of the animal, and climatic conditions. Meat fats also contain small portions of other minor fatty acids, which we have not included in this discussion.

Fish oil/marine oil is a versatile product and finds many applications in the food, feed, and technical industries of the world. The largest use for fish oil is in its partially hydrogenated form in Europe

Table 2.5. **Fatty Acid Composition of Some Meat Food Fats**

		% Butterfat	% Lard	% Beef Tallow
C_4	Butyric	4	—	—
C_6	Caproic	3	—	—
C_8	Caprylic	1	—	—
C_{10}	Capric	2	—	Trace
C_{12}	Lauric	3	—	Trace
C_{14}	Myristic	12	2	4
C_{16}	Palmitic	26	25	26
C_{18}	Stearic	13	13	22
$C_{18:1}$	Oleic	29	45	39
$C_{18:2}$	Linoleic	3	10	3
$C_{18:3}$	Linolenic	1	Trace	1
C_{20}	Arachidic	—	Trace	Trace
$C_{20:4}$	Arachidonic	Trace	1	1

Table 2.6. Principal Fatty Acids of Some Commercially Available Marine Oils, as a Percentage of Oil

	Menhaden	Herring	Anchovy	Cod Liver	Mackerel	Sandee Mackerel
$C_{14:0}$	9	7	9	3	8	7
$C_{16:0}$	20	16	19	13	14	15
$C_{16:1}$	12	6	9	10	7	8
$C_{18:1}$	11	13	13	23	13	9
$C_{20:1}$	1	13	5	0	12	15
$C_{22:1}$	0.2	20	2	6	15	16
$C_{20:5}$	14	5	17	11	7	9
$C_{22:6}$	8	6	9	12	8	9

where it is used in cooking and baking fats. The U.S. Food and Drug Administration approved partially hydrogenated and hydrogenated menhaden oils as GRAS (generally recognized as safe) on September 15, 1989 (2). The U.S. market is developing slowly, so the United States continues to export about 80% of its production to Europe for edible use. The remaining 20% is used in animal feeds and in a number of different industrial applications.

Marine oils consist of a mixture of triglycerides of various long-chain fatty acids with small amounts of monoglycerides, diglycerides, free fatty acids, and sterols. The fatty acids that characterize marine oils are similar to those in the various vegetable oils and animal fats differing principally in their high proportion of long-chain polyunsaturated fatty acids with five and six double bonds. The principal marine oil polyunsaturated fatty acids are in the omega-3 position, and the major vegetable oil polyunsaturates are in the omega-6 position.

Marine oils differ among themselves in the percentage of these fatty acids. Table 2.6 gives a comparison of the composition of the major fish oils of commerce (5–7).

SUMMARY

This chapter covered the basic organic chemistry/structure of oils and fats. Structures were illustrated and described for the glycerol molecule: triglycerides (simple and mixed); diglycerides; monoglycerides; carboxyl group; saturated, unsaturated, and polyunsaturated fatty acids; and conjugated and nonconjugated fatty acids. The most common food fatty acids were tabulated using their common names as well as using the omega and Geneva methods of classifying them. Typical compositions of some important vegetable oils and meat fat were also tabulated for easy reference.

References

1. Swern, D. 1979. *Bailey's Industrial Oil and Fat Products, Vol. 1*, 4th ed. New York: Wiley Interscience, p. 16.
2. "A Changing Future for Fish Oils," *Oils & Fats International Issue Five*, p. 20, 1992.
3. Inst. of Shortening Edible Oils. 1988. *Food Fats and Oils*, 6th ed. Institute of Shortening and Edible Oils, Washington, DC: p. 4.
4. Swern, D. 1979. *Bailey's Industrial Oil and Fat Products, Vol. 1*, 4th ed. New York: Wiley-Interscience, pp. 292–322 and 332–350.
5. Swern, D. 1979. *Bailey's Industrial Oil and Fat Products, Vol. 1*, 4th ed. New York: Wiley-Interscience, pp. 448–452.
6. Kinsella, J. 1987. *Seafoods and Fish Oils in Human Health and Disease*. New York: Marcel Dekker, pp. 66–67, 114–115, 211, and 260–261.
7. Barlow, S. 1982. *Nutritional Evaluation of Long-Chain Fatty Acids in Fish Oil*. London: Academic Press, pp. 25–31, 47–49, 38–39, 44, and 50–54.

General References

Hui, Y. 1992. *Encyclopedia of Food Science and Technology, Vol. 2*, New York: John Wiley and Sons.
Fieser, L. 1961. *Advanced Organic Chemistry*. New York: Reinhold.

Chapter 3

Common Chemical Reactions

Knowledge of the important chemical changes that oils and fats may undergo is necessary to understand how the various products are manufactured as well as to cope with the possible problems that can occur in storage, transportation, and use. The most important chemical reactions occur at (1) the points of unsaturation on the fatty acid chain and (2) the point where the fatty acids are attached to the glycerol molecule (the ester linkage). This knowledge will be especially helpful in understanding the changes that take place in fats used in deep-frying.

HYDROLYSIS

Hydrolysis is the reaction of water with a substance, such as fats. This results in the splitting of some of the fatty acids from the oil or fat, yielding some free fatty acids. Some monoglycerides and di-glycerides are produced, but in the frying operation this is not significant. The small amounts produced will be distilled from the hot frying fat. In addition, in some situations this partial hydrolysis

yielding some monoglycerides and diglycerides will eventually be
carried to completion, resulting in glycerol and free fatty acids.

$$
\begin{array}{l}
\text{H} \\
| \\
\text{H—C—O-Fatty acid radical} \\
| \\
\text{H—C—O-Fatty acid radical} \\
| \\
\text{H—C—O-Fatty acid radical} \\
| \\
\text{H}
\end{array}
\quad +3\text{HOH} \xrightarrow{\text{Heat}}
\begin{array}{l}
\text{H} \\
| \\
\text{H—C—OH} \\
| \\
\text{H—C—OH} \\
| \\
\text{H—C—OH} \\
| \\
\text{H}
\end{array}
\quad + 3 \text{ Free fatty acids}
$$

Triglyceride   + Water $\xrightarrow{\text{Heat}}$ Glycerol + 3 Free fatty acids

In the digestive tracts of humans and animals, fats are hydrolyzed
by lipase enzymes. This is part of the digestion process and will be
covered in a later chapter.

Hydrolysis is a reaction that takes place at the junction of the fatty
acids and the glycerol portion of the molecule. Hydrolysis is accelerated by high temperatures and pressures and an excessive amount
of water. This reaction is especially significant in the preparation of
deep-fried foods, where the frying fat may be at a temperature of
350°F (176.6°C) and the food that is fried is high in moisture. A good
example of this situation is in the frying of french fried potatoes.
Fresh potatoes are high in water, over 80% of their weight before
cooking, and during frying free fatty acids can develop at a fairly
rapid rate. Moderate levels of free fatty acids, up to 3% in foodservice frying and up to 1% in large commercial frying, do not necessarily have an adverse effect on frying fat performance. However,
excessively high levels may result in excessive smoking and even
affect the flavor of the fried food.

HYDROGENATION

This is one of the more important chemical reactions of food oils
and fats, especially oils. It is a typical example of a reaction that
occurs at points of unsaturation or double bonds.

$$
\begin{array}{l}
\text{H} \;\; \text{H} \;\; \text{O} \\
| \quad | \quad \| \\
\text{R—C}=\text{C—C—OH} + \text{H}_2
\end{array}
\xrightarrow[\text{Nickel}]{\text{Heat}}
\begin{array}{l}
\text{H} \;\; \text{H} \;\; \text{O} \\
| \quad | \quad \| \\
\text{R—C—C—C—OH} \\
| \quad | \\
\text{H} \;\; \text{H}
\end{array}
$$
catalyst

Fatty acid radical + Hydrogen   Hydrogenated fatty acid

Hydrogenation Reaction

FIGURE 3.1. Hydrogenation of soybean oil. Clockwise from left—light hydrogenation, fluid shortening, plastic shortening, and complete hydrogenation.

Hydrogen is added directly to points of unsaturation in the fatty acids. Gaseous hydrogen is used at elevated oil temperatures under increased pressure and in the presence of a suitable catalyst such as a nickel compound. The catalyst speeds up the desired reaction. The catalyst does not react itself, and it is removed from the fat after the hydrogenation is completed. This reaction is used to make fat products with greater flavor stability, minimizing the possibility of oxidation, especially if fatty acids such as linolenic (three points of unsaturation on the same fatty acid) are originally present. Linolenic acid, which is present at levels of about 8% in soybean oil, would oxidize readily and produce rancid flavors if the oil were not lightly hydrogenated. Hydrogenation also permits the conversion of liquid vegetable oils into fluid shortenings and semisolid plastic shortenings that are better adapted for use in deep-frying, baking, and so on, because the addition of hydrogen will raise the melting point.

The importance of hydrogenation is further understood when we recognize that soybean oil now provides about 60–65% of the total visible fat in the U.S. diet (1).

Hydrogenation is a reaction used to optimize the properties of fats and oils required for specific uses. The hydrogenation reaction is easily controlled and can be stopped at any point. Usually a variety of stocks of different degrees of hydrogenation are made, from the very lightly hydrogenated oils, to oils with intermediate degrees of hydrogenation, and to completely hydrogenated stocks (Fig. 3.1). Various stocks are blended to obtain the desired properties in the finished shortening or oil.

The rate of hydrogenation depends on the following:

1. Nature of substance to be hydrogenated. The more the double bonds, the faster the rate.

2. Nature and concentration of catalyst. In general, an increase in concentration will increase the reaction rate, increase selectivity (between $C_{18:3}$ and $C_{18:2}$ for example), and increase trans acid formation. The most important catalysts contain nickel or copper. The main advantages for nickel include (1) availability, (2) low cost, and (3) inert nature of the metal to oil. The main disadvantage for nickel is the difficulty in hydrogenating selectively along with low trans acid formation. Copper catalysts provide greater selectivity and less trans acid formation as compared with nickel types. On the other hand, copper is less active than nickel, its reuse is not practical, its consumption is 5–10 times that of nickel, and copper is a pro-oxidant and requires considerable posttreatment to eliminate all traces in the hydrogenated product.

3. Concentration of hydrogen. Increases in hydrogen will increase the reaction rate.

4. Reaction temperature. Increased temperature will increase reaction rate, selectivity, and trans acid formation.

5. Pressure. In general, an increase in pressure will increase reaction rate, reduce selectivity, and reduce trans acid formation.

6. Agitation. Increased agitation will also increase reaction rate, reduce selectivity, and reduce trans acid formation.

OXIDATION

Like hydrogenation, this reaction occurs at the double bonds or points of unsaturation.

$$\begin{array}{ccc} \text{H} & \text{H} & \text{O} \\ | & | & \| \\ \text{R--C}\!=\!\text{C--C--OH} + \text{O}_2 \end{array} \xrightarrow[\text{Time}]{\text{Light, heat}} \begin{array}{ccc} \text{H} & \text{H} & \text{O} \\ | & | & \| \\ \text{R--C--C--C--OH} \\ & | & | \\ & \text{O--0} \end{array}$$

Fatty acid radical + Oxygen → Peroxide

Oxidation Reaction

This is the reaction of an oil or fat with oxygen in the air, and with food this is not desirable because the reaction will adversely affect the flavor of the fat and the food in which it is used. In fact, considerable care is exercised during manufacturing, storage, and usage to either keep this reaction from occurring or slow it down as much as possible.

Oxidation induced by air at room temperature is referred to as autoxidation. Generally, this is a slow process; considerable time is needed to produce a sufficient quantity of peroxides (the main initial products of autoxidation) to develop objectionable flavors and odors.

Products containing a higher proportion of unsaturated fatty acids are more prone to oxidation than those containing lesser amounts.

The rate of oxidation increases with an increase in temperature, exposure to oxygen in the air, the presence of light, and contact with materials that are classified as pro-oxidants. An excellent example of a pro-oxidant is the metal copper. Therefore, care should be taken to keep copper, brass, bronze, and other copper-containing alloys out of oil and fat processing systems, their packages, and food-manufacturing plants that utilize oils or fats. In deep-frying, where the temperature of the fat is high, it is important to keep copper from coming in contact with fat, especially at the surface, where the fat is also in contact with oxygen in the air. Examples of sources of copper contamination in deep-frying operations are copper thermocouples and brass or bronze drain valves. Copper welds to repair baskets are another source of copper contamination.

Natural oils and fats from vegetable sources contain minute amounts of substances capable of inhibiting oxidation to a certain extent. Alpha tocopherol is the most important and almost universally distributed natural antioxidant.

Lard, hydrogenated lards, tallow, and shortenings made from meat fats are often fortified or stabilized by the addition of antioxidants, since they do not contain natural antioxidants. They are generally added at a very low level, such as a few parts per million. Two of

Table 3.1.   **Some Direct Food Additives Used in Fats and Oils**

Additive	Effect Provided
Tocopherols	Antioxidant, retards oxidative rancidity
Butylated hydroxyanisole (BHA)	Antioxidant, retards oxidative rancidity
Butylated hydroxytoluene (BHT)	Antioxidant, retards oxidative rancidity
Tertiary butylhydroquinone (TBHQ)	Antioxidant, retards oxidative rancidity
Carotene (pro-vitamin A)	Color additive, enhances color of finished foods
Methyl silicone (dimethylpolysiloxane)	Inhibits oxidation tendency and foaming of fats and oils during frying
Diacetyl	Provides buttery odor and flavor to fats and oils
Lecithin	Water scavenger to prevent lipolytic rancidity
Citric acid $\Big\}$ Phosphoric acid	Metal chelating agents, inhibit metal-catalyzed oxidative breakdown

Courtesy of Institute of Shortening and Edible Oils.

the commercially more important ones are butylated hydroxyanisole (BHA) and tertiary butylhydroquinone (TBHQ).

These low-temperature antioxidants do not provide much protection to fats and oils at frying temperatures because of their tendency to distill off.

Some of the materials added to fats and oils are summarized in Table 3.1.

Even some vegetable shortenings and oils that are processed under questionable standards may include added antioxidants. This is done to slow down deterioration due to poor processing conditions.

Under normal circumstances, the oxidation of fats and oils is a slow process. Slight degrees of oxidation will not be noticed by most users, but after a time—usually measured in weeks or months, depending on storage conditions and the stability of the product—oxidation can proceed to the point where a rancid flavor and/or odor is noticed.

In the frying kettle, another reaction occurs that is related to oxidation—fat color darkening. Trace elements from the food may react with a portion of the frying fat to cause the frying fat to darken. This can have an adverse effect on the color and appearance of fried food.

In scientific investigations, measurement of the degree of oxidation is usually performed by measuring the amount of oxygen absorbed or the changes in the peroxide value of a given weight of fat with time. During the initial phase of oxidation, the changes are relatively slow and occur at a more or less uniform rate. This is called the induction period. After a certain amount of oxidation occurs, this reaction reaches a second phase in which oxidation may occur at a very rapid rate as fats and oils develop rancid flavors and odors.

It must be pointed out that the measurement of peroxide value has a very limited value in determining the condition of a frying fat. At frying temperatures (about 350°F or 176.6°C), these peroxides are quite volatile and they are continuously removed by distillation.

In a further advanced study of oxidative changes, the free-radical theory of oxidation includes a three-stage series of reactions. The following is a simplified version of the changes that occur:

1. Initiation: The fat molecules form fatty free radicals in the presence of initiators such as ultraviolet light, heat, and heavy metals such as copper.

$$\underset{\text{Fat molecule}}{R\,H} \xrightarrow{\text{Initiators}} \underset{\text{Fatty free radical}}{R + H}$$

2. Propagation:

$$R + O_2 \rightarrow \underset{\substack{\text{Peroxide} \\ \text{free radical}}}{R\,O\,O}$$

$$R\,O\,O + R\,H \rightarrow \underset{\text{Hydroperoxide}}{R\,O\,O\,H + R}$$

The decomposition of hydroperoxides leads to the formation of a wide variety of aldehydes, ketones, and hydrocarbons. These materials are responsible for the rancid odors and flavors.

3. Termination of the oxidation chain reaction occurs if the free radicals are deactivated or destroyed. Oil and fat antioxidants such as tocopherols can react with the initiating and propagating radicals to produce harmless products.

POLYMERIZATION

This is the reaction of a fat with itself, whereby relatively small molecules of oil or fat combine to form much larger molecules. Polymerization may occur either at points of unsaturation on fatty acid chains (preceded by oxidation) or at the juncture of the fatty acid and the glycerol molecule.

$$R + R \rightarrow {}^2(R)$$

Polymerization Reaction

The polymerized molecule may be as much as hundreds or thousands of times the molecular weight of the original molecules.

Polymerization can occur in the deep frying of foods, where frying is done at temperatures ranging from 325°F (162.8°C) to 375°F (190.6°C). The reaction is accelerated by frying at too high a temperature (above 350°F, 176.6°C), the presence of oxygen, the use of a poor-quality frying fat, and poor frying practice—for example, heating for long periods of time with little or no frying of foods. (This practice reduces the rate at which fat is removed from the kettle by absorption into the food and, in turn, reduces the rate at which fresh fat is added to the system.)

All commonly used food fats, and particularly those high in polyunsaturated fatty acids, tend to form these polymers when heated at extreme temperatures and for an extreme length of time.

Polymerization is evidenced by deposits of a gumlike material around the sides of the frying kettle, frying baskets, commercial frying kettle conveyors, and so on, especially where fat, metal, and air are in contact with each other. Polymerization also causes an increase in the viscosity of the frying oil. If polymerization is allowed to proceed to extremes, it can result in foaming of the frying fat. When the frying fat is still usable, the water in the foods being fried evolves as large bubbles and is quickly driven out of the fry kettle. However, when this frying fat polymerizes to a considerable extent and its viscosity increases, the moisture in the food evolves as very small bubbles during frying and rises slowly up the sides of the frying kettle. This is referred to as foaming. The extreme difference in the sizes of normal fat molecules and the polymerized molecules is an important factor in this phenomenon.

The use of the proper type and amount of methyl silicone in a frying fat, added at the proper point in processing, helps to retard the development of foam. When foaming occurs, the entire contents of the kettle must be discarded and replaced with fresh shortening.

In general, the rate of polymerization increases with the amount of unsaturation in a fat or oil. However, by utilizing the latest processing techniques, including partial and very selective hydrogenation of soybean oil along with addition of the correct amount and type of methyl silicone, frying fats relatively high in polyunsaturation can be made to resist foaming.

ESTERIFICATION

In its simplest form, esterification may be considered to be the reverse of hydrolysis. It is the combining or recombination of free fatty acids with glycerol to form triglycerides. Monoglycerides and diglycerides may also be produced by esterification.

An important commercial reaction involves the production of monoglycerides and diglycerides from triglycerides and glycerol. This reaction is sometimes referred to as glycerolysis or superglycerination. See Fig. 3.2. Monoglycerides are important as emulsifying agents in many food products. An emulsifier tends to hold fat and water together (they would normally separate). An example is in the mixing of cake batters. The OH portion of the molecule tends to hold onto the water and water-soluble portions of the batter (e.g., water and sugar). The fatty acid portion of the monoglyceride tends to latch onto the fat-soluble portions of the batter (e.g., shortening and egg yolk). A smooth, nonseparating cake batter results in a baked cake of finer grain and smoother texture.

INTERESTERIFICATION

This reaction can be explained as a migration and interchange of fatty acid radicals from one fat to another or from one point to another. This is done to develop new fat molecules that have specific properties. Monoglycerides and diglycerides may also be produced by certain interesterification reactions. Interesterification or ester interchange (rearrangement) reactions may also be subdivided into either random interesterification or directed interesterification.

In the case of random rearrangement, the ester interchange may proceed at random, with the eventual attainment of an equilibrium composition of products corresponding to the laws of probability. This reaction may be carried out at very high temperatures (480°F

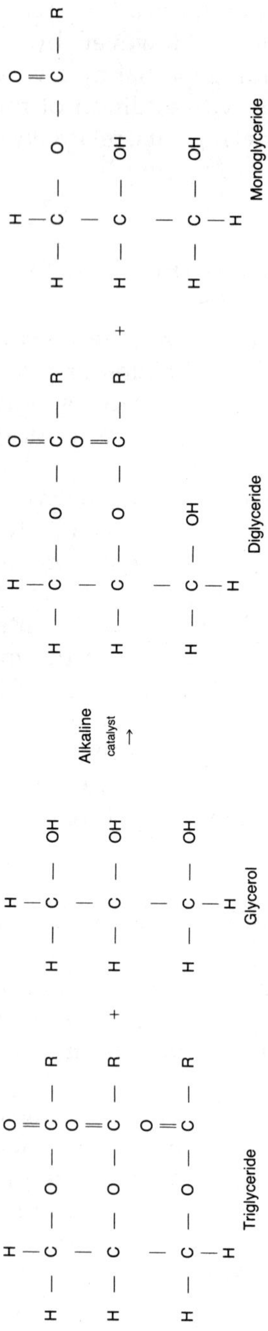

FIGURE 3.2.   Esterification reaction.

or 249°C or higher) without the use of a catalyst. More likely, the reaction is carried at much lower temperatures along with the use of an alkali metal catalyst.

In the case of directed rearrangement, the reaction is generally carried out at temperatures of around 90–100°F or 32.2–37.8°C in the presence of a catalyst such as sodium methylate. Under these reaction conditions, it is possible to continuously remove trisaturated glycerides by crystallization. This enables the production of a shortening with a better plastic range.

A commercial example of interesterification is the directed rearrangement of lard to enable lard to perform more like vegetable shortening. Natural lard from the pig tends to produce very large coarse crystals in the beta polymorphic form (see Chapter 4). Directed interesterification enables lard to crystallize in the beta prime form. This type of processing tends to improve melting point, plastic range, and crystal structure.

The rearrangement process does not change the degree of unsaturation of the fatty acids because they transfer in their entirety from one position or one molecule to another (2–4).

HALOGENATION

The halogens include chlorine, bromine, and iodine. They can readily add to the double bonds of unsaturated fatty acids as follows:

$$\begin{array}{ccc} H & H & O \\ | & | & \| \\ R-C=C-C-OH + I_2 \end{array} \qquad \begin{array}{ccc} H & H & O \\ | & | & \| \\ R-C-C-C-OH \\ | & | & \\ I & I & \end{array}$$

Measured quantities of iodine monochloride may be added to measured quantities of fats or oils to determine the average degree of unsaturation of this fat or oil. This, in turn, results in the iodine number (or iodine value), an important analytical measurement. This will be discussed further in Chapter 14.

ISOMERIZATION

Isomers are two or more substances that are composed of the same elements combined in the same proportions, hence having the same molecular formula but differing in molecular structure. The two important types of isomerism among fatty acids are (1) geometric and (2) positional.

Geometric Isomerism

Unsaturated fatty acids can exist in either the cis or trans form depending on the configuration of the hydrogen atoms attached to the carbon atoms joined by the double bonds. If the hydrogen atoms are on the same side of the carbon chain, the arrangement is called cis, and if the hydrogen atoms are on opposite sides of the carbon chain, the arrangement is called trans, as shown by the following diagrams:

$$
\begin{array}{cc}
\overset{\displaystyle H\ \ H\ \ H\ \ H}{\underset{\displaystyle H\qquad\quad H}{-C-C=C-C-}} & \overset{\displaystyle H\ \ H\qquad H}{\underset{\displaystyle H\quad\ H\ H}{-C-C=C-C-}} \\
\text{cis} & \text{trans}
\end{array}
$$

Elaidic and oleic acids are examples of geometric isomers; both have the same molecular formula but are geometric isomers. In the case of elaidic, the double bond is in the trans configuration, but for oleic, the double bond is in the cis configuration.

Positional Isomerism

In this case, the location of the double bond along the fatty acid chain differs among the isomers.

The position of the double bonds affects the melting point of the fatty acid to a limited extent. Processing such as hydrogenation can cause shifts in the location of double bonds in the fatty acid chains as well as cis–trans isomerization.

The possible number of positional and geometric isomers increases with the number of double bonds. For example, with two double bonds, the following four geometric isomers are possible: cis–cis, cis–trans, trans–cis, and trans–trans. Trans–trans are present in only trace amounts in partially hydrogenated fats and, thus, are insignificant in the human food supply. The geometric configuration has an appreciable effect on the melting point of the fatty acid.

Generally speaking, cis isomers are those naturally occurring in food fats and oils, although small amounts of trans isomers occur in fats from ruminants. Most trans isomers result from the partial hydrogenation of fats and oils.

SUMMARY

The most significant reactions encountered in oil and fats are hydrolysis, hydrogenation, oxidation, polymerization, esterification, interesterification, and isomerization. These are important in manufacturing oil and fat products from their naturally occurring basic vegetable oils and meat fats, in changes that occur during storage and usage, and in their further processing into foods that are ready for sale to the consumer.

References

1. Institute of Shortening Edible Oils, 1988. *Food Fats and Oils*, 6th ed. Washington, DC: Institute of Shortening and Edible Oils, p. 19.

2. Rozenaal, A. 1992. ·Inform *3*(11), 1232–1237.

3. Eckey, E. 1948. Ind. Eng. Chem. *40*, 1183.

4. Going, L. 1967. U. Amer. Oil Chem. Soc. *44*, 414A.

Chapter 4

Physical Properties

The physical properties of oils and fats are of practical importance in understanding the makeup of these materials and how they should be used.

GENERAL OVERVIEW

The physical characteristics of an oil or fat are dependent on such factors as seed or plant source, degree of unsaturation, length of carbon chains, isomeric forms of the fatty acids, molecular structure of the glycerides, and processing.

Fats that are liquid at room temperature tend to be more unsaturated than those that appear to be solid. The degree of unsaturation can be expressed in terms of the iodine value of the fat. Iodine value is defined as the number of grams of iodine that will react with the double bonds in 100 g of fat. The higher the iodine value, the more the unsaturation of a specific oil or fat. The typical iodine value of crude soybean oil is in the 125–135 range. A processed salad oil will have an iodine value of about 110–115. A typical semisolid shortening made from partially hydrogenated vegetable oils will have an iodine value of 85–95. (Refer to the discussion of solids content index in the subsection "Melting Points" as well as Table 4.1.)

As the chain length of the fatty acid increases, the melting point also increases. Thus, a short-chain saturated fatty acid such as butyric acid will have a lower melting point than saturated fatty acids with longer chains and even some of the higher-molecular-weight unsaturated fatty acids, such as oleic acid. This explains why coconut oil, which is almost 90% saturated fatty acids but has a high proportion of short-chain fatty acids, is liquid, whereas lard, which contains only about 37% saturates, most with longer chains, is solid-appearing at about 78°F (25.6°C). (See the discussion of solids content index later in this chapter.)

For a given fatty acid chain length, saturated fatty acids will have higher melting points than those that are unsaturated, but this generalization sometimes is complicated by the presence of geometric isomers of the unsaturated fatty acids. For example, oleic acid (cis) and its geometric isomer elaidic acid (trans) do not have the same melting point. Oleic acid is liquid at temperatures considerably below room temperature, whereas elaidic acid is solid even at temperatures considerably above room temperature. Thus, the presence of different geometric isomers of fatty acids influences the physical characteristics of the fat. Elaidic, the trans acid, is more linear and will pack better in a crystalline array and will, therefore, melt at a higher temperature.

The molecular structure of triglycerides can also affect the properties of an oil or fat. A single triglyceride will have a sharp melting point. A mixture of triglycerides, as is typical of lard and finished vegetable oil and fat products, will have a broad melting range.

A mixture of several triglycerides has a melting point lower than the average of the melting points of the individual components. The mixture also has a broader melting range than any of its components. Monoglycerides and diglycerides have higher melting points than triglycerides with identical fatty acid compositions.

Solidified fats can exist in several different crystalline forms, depending on the manner in which the molecules orient themselves in the solid state. This crystal structure can vary with the source oil, manufacturing and tempering procedures, and storage conditions. See the discussion regarding polymorphism later in this chapter as well as Chapter 6 (Processing Technology).

FLAVOR

Almost all oils or fats consumed in the United States are preferred to be either as bland in flavor as possible or to have a "butter-like" flavor. This preference depends on the intended use of the product.

Butterfat has a distinctive and desirable flavor and aroma. However, if butterfat is permitted to become rancid, its flavor and aroma are no longer desirable and might be considered objectionable or distasteful by many consumers. The same is true for margarine or any other oil or fat.

The flavor of lard was at one time considered to be desirable in the United States, but today blandness or absence of flavor in shortenings is much preferred. The normal flavor of lard is still considered desirable in some parts of Europe. Unrefined (virgin) olive oil also has a distinct flavor that is prized quite highly by people living in areas surrounding the Mediterranean Sea and by people in the United States whose families were originally from countries in the Mediterranean area, including Spain and Italy. In addition, the sales of olive oil in the United States have doubled over the past 5 years because of its perceived nutritive value.

Crude vegetable oils such as cottonseed, soybean, and palm oils have distinctive undesirable flavors, but these oils are processed to attain the desired bland, neutral flavor. For further specifics on flavor, please refer to chapter 16.

MELTING POINTS

The complete melting point is the temperature at which a solid fat becomes a liquid oil. Each individual pure fatty acid has a specific complete melting point. As oils and fats are essentially mixtures of various fatty acids as triglycerides (e.g., stearic, oleic, linoleic, etc.), these oil and fat products do not have sharp melting points. For example, as the temperature of a shortening rises, some fatty glycerides melt, and as the temperature drops, some portions of this product resolidify. The amount of the fat that is solid at a given temperature can be determined analytically.

The complete melting point of a specific oil or fat is the temperature at which that particular product is completely melted. For vegetable shortenings, the complete melting point is about 120°F (49°C). Fluid shortenings also have a complete melting point of about 120°F (49°C). The complete melting point for a specific product may be quite misleading in a study of its physical properties. More pertinent is knowledge of the ratio of solids to liquids at temperatures

Table 4.1. Solid Fat Index

Complete Melting Point	Solid Shortening 119–121°F (48.2–49.3°C)	Fluid Shortening 119–121°F (48.2–49.3°C)
Solids content at		
50°F (10°C)	23%	8%
70°F (21°C)	18%	7%
92°F (33°C)	15%	7%
105°F (40.5°C)	9%	6%
120°F (49°C)	0%	0%

from 50 to 120°F (10–49°C). Sometimes this is expressed as the solid fat index or solids content index.

The comparison of the solids content at various temperatures for a solid shortening and a fluid shortening will help illustrate this point. See Table 4.1.

The following factors are important in determining the complete melting point and the melting behavior of a product:

1. The average chain length of the fatty acids. In general, the longer the average chain length, the higher the melting point.

2. The positioning of the fatty acids on the glycerol molecule also affects the melting point. As an example, safflower oil, which has a long average chain length, will melt like a medium chain length triglyceride.

3. The relative proportion of saturated to unsaturated fatty acids. The higher the proportion of unsaturated fatty acids, the lower the melting point.

4. Processing techniques such as the degree and selectivity of hydrogenation and winterization. These are discussed in Chapter 6.

PLASTICITY

From the discussion on melting point, it was seen that a shortening that appears solid to the eye at room temperature is really composed of both solid fats and liquid oils. At 70°F (21°C), a typical shortening that appears solid contains 15–20% solids and, hence, 80–85% liquid oil. Through proper selection of oil/oils, proper processing techniques, and storage control, this small amount of solids can be made to hold all of the liquid in a matrix of a very small, stable, needlelike crystals (beta-prime crystals). The single most important factor in developing beta-prime crystals is the use of some strong beta-prime tending oil (such as hydrogenated palm oil) in the formula. As the temperature drops, more of the oils solidify and the shortening becomes progressively firmer. On the other hand, as the temperature rises, more solid fats or fatty acids melt and the shortening becomes progressively softer until it has practically no "body" or plasticity at all, and eventually it becomes completely melted.

If a shortening that is workable over a wide temperature range is desired, then it should be made up of a combination of triglycerides ranging widely in melting points. By the same reasoning, when a fat or shortening with a narrow temperature range of workability is needed, it is made up of a greater amount of triglycerides of similar melting points.

POLYMORPHISM

The section on plasticizing in the chapter on processing (Chapter 6) touches on polymorphism. Polymorphism is the ability of fat crystals to exist in more than one crystal form or modification. These crystal forms are alpha, beta-prime, and beta, in order of increasing stability. The change is said to be monotropic, that is, always proceeding in the solid phase from lower to higher stability. The forms differ in crystalline structure and in melting points. Both are of appreciable importance in fat products that have solid contents at use temperatures such as shortenings (plastic) and fluid shortenings, which are pourable and pumpable.

Some investigators believe there are more than three polymorphic forms for glycerides. An example is an intermediate polymorph between beta-prime and beta, said to have large crystals. Others say a less dense reversible sub-alpha form also exists. Neither the intermediate nor the sub-alpha is believed to have any practical significance.

The fatty acid makeup of and position in the glycerides of the fat solids and temperature history are the two main factors in determining polymorphic behavior. Other factors include kind and quantity of impurities, nature of possible solvent, and degree of supercooling. A high level of fatty acids of identical chain length results in a slow conversion rate of beta-prime to beta and a coarsening of crystal structure. The more heterogenous the fatty acid makeup, the more likely it will be beta-prime and fine-grained or needlelike crystals.

X-ray diffraction patterns are used to identify the polymorphic forms. Alpha short spacings are found at 4.15Å, 4.2Å, and 3.8Å, and beta at 4.6Å, 3.8Å, and 3.7Å (1).

Fat crystal sizes usually run 1–10 μm. Alphas are fragile transparent platelets of about 5 μm. Beta-prime are tiny needles of about 1 μm length. Beta are large, coarse crystals of about 25–50 μm. The intermediate form is said to be 3–5 μm in size.

If a fat is cooled rapidly, the tendency is to form the small, alpha crystals. These generally do not last long and convert rapidly to the beta-prime needlelike crystals. These beta-prime crystals are considered highly stiffening and, hence, are the form of choice for plastic shortenings. Depending on the glyceride makeup and the temperature history, the beta-prime form may convert to the most stable beta form. Beta has large, coarse, platelike crystals. These are not stiffening; hence, those hydrogenated fats exhibiting this behavior are the choice for the solids in fluid shortenings. In fact, the discovery of fluid shortenings resulted from a happenstance that occurred around 1950. This and two other commercial developments of that same time period will serve to illustrate the polymorphism of component solids in the physical behavior of the resulting commercial products.

In the 1940s the major shortening products were cottonseed oil products, in both the base stock and hardstock, and were beta-prime (solids). Later in the 1940s, a less expensive soybean oil appeared on the market but was somewhat suspect flavorwise because of its high content of linolenic acid. It was thought that soybean oil products would revert to a beany flavor. However, for cost reasons, the manufacturer of one leading brand decided to replace the cottonseed oil hardstock by bean. The plasticized product met all specifications. However, when exposed to a particular temperature history, this plastic product became pourable. Consumer performance in baking, frying, and so on was unimpaired, but it looked and handled differently. It was indeed a new product. However, because of the "soupy" consistency, this product with the soybean oil hardstock soon lost its brand franchise. Further investigations proved that (1) in making a plastic shortening it was necessary to use a beta-prime tending hardstock such as cottonseed oil hardstock and that (2) by learning from this mistake, fluid shortenings could be produced.

Another development around 1950 was to use the propensity of peanut oil hardstock to have good stability in the beta-prime form to market a peanut butter stabilizer (prevent free oil separation on the shelf) composed of one-third each of peanut oil, peanut oil hardstock, and salt. As manufactured, the stabilizer was in the beta form, hence pumpable (fluid). The peanut butter manufacturers could easily handle it in formulation and processing. When solidified or plasticized, the hardstock crystallized in the beta-prime form. The result was no free oil separation in the finished butter.

The last commercial example deals with the alpha form, the form that tends to be waxy and translucent. In the late 1940s it was discovered that the triglycerides with two acetyl radicals and one high-molecular-weight saturated acid like stearic were stable in the alpha form—stable for years. These materials were also unique in having sharp melting points near body temperature. These acetin fats or acetoglycerides found worldwide application as coating materials for foods such as meats and cheeses.

Generally speaking, beta-prime forms melt about 5–10°C (9–18°F) higher than the alpha forms, and the betas also melt about 5–10°C (9–18°F) above the beta-prime forms. As illustrations, we note that alpha tristearin melts at 54.7°C (130.5°F), beta-prime tristearin at 64.0°C (147.0°F), and beta tristearin at 73.3°C (163.9°F).

FLUIDITY OF FLUID SHORTENINGS

The techniques of manufacturing fluid shortenings will be discussed in Chapter 6. Fluid shortenings require the development of large "ball-bearing-type" crystals (beta) that provide ease of movement or fluidity to give pourability and pumpability over the range 50°–100°F (10–38°C). This beta type of crystal structure is sometimes referred to as platelets and is the most stable of the three key polymorphic forms.

COLOR

Whiteness of color is generally preferred, except with butter, margarine, and some liquid and plastic shortenings that are intentionally given an added yellow color, usually obtained through the use of minute quantities of beta-carotene. It is important to use the correct level of approved yellow color such as beta-carotene or annatto. Too high a level may result in an unappetizing product. In addition, these yellow coloring materials must be added near the end of processing in order to minimize their loss.

OILINESS

Fatty materials feel oily and they have the ability to form oily or lubricating films. In preparing some food products, such as grilled foods, this lubricating action is very important.

VISCOSITY

Viscosity is a measure of internal friction between molecules. In general, the viscosity of oils decreases slightly with an increase in unsaturation; therefore, viscosity is increased slightly by hydrogenation. Oils or fats containing a greater proportion of fatty acids of relatively low molecular weight are slightly less viscous than ones of an equivalent degree of unsaturation but containing a higher proportion of high-molecular-weight acids. The viscosity of highly polymerized oils is much greater than that of normal oils.

Viscosity is occasionally referred to in determining the condition of fats used in deep frying. During use in the frying kettle, the viscosity of a frying fat or oil will tend to increase as oxidation and polymerization increase. This can be related to polymer development and tendency toward foaming.

EMULSIFICATION

Triglycerides with three fatty acids attached to a glycerol molecule have minimal emulsification properties. However, monoglycerides with only one fatty acid attached to a glycerol molecule with two free hydroxyl groups (OH) on the glycerol molecule take on some of the properties of both fats and water. The fatty acid portion of the molecule acts like any other fat and readily mixes with these fatty materials, whereas the two OH groups mix or dissolve in water; thus monoglycerides tend to hold fats and water together. This is especially important in mixing cake batter, where it is necessary to hold the shortening and water (including the liquids in milk and eggs) together to form a stable batter that will not readily separate.

With the addition of monoglycerides to all-purpose shortenings, high ratio shortenings evolved. The term high ratio came about because the finer dispersion of this type of shortening in cake batters permitted the professional baker to use higher ratios of sugar (as well as shortening, milk, and eggs) in relation to flour, as compared with regular or all-purpose shortenings. This permitted the baker to produce cakes that were moister, of better flavor and eating qualities, and better keeping qualities. The monoglyceride level of high-ratio shortenings for household use is generally in the 1.5–2.0% range, and for commercial baking in the 2.5–3.5% range. This same type of high-ratio shortening also resulted in icings of better volume, texture, and eating quality.

These same emulsifiers also act as softeners (retard crumb firming) in yeast goods such as bread, rolls, sweet yeast-raised pastry, and Danish pastry. In special bread and sweet dough shortenings for commercial bakery use, the monoglyceride level is even higher, in the 6.0–8.0% range.

Other "formulated" emulsifiers have entered the food field, generally to fit a specific and limited function. These special emulsifiers are most useful in prepared cake mixes, providing quick aeration or foaming ability when the prepared mix cake batters are mixed in the home. Examples are lactic acid esters of monoglycerides and diglycerides, polyoxyethylene sorbitan monosteareate, and propylene glycol esters of fatty acids (PGMS/PGMP). These emulsifiers work well in providing cakes that are very moist and tender. Polyoxyethylene sorbitan monostearate (e.g., PS60) is used to a certain extent in "icing shortenings." It provides good volume or aeration in lean (low fat-to-sugar ratio) wholesale-type cream icings, but it deteriorates rapidly, especially in the summer, making it generally undesirable for use in rich icings, in decorator icings, or in hot-weather areas. See Table 4.2 for a list of some of the more commonly used food emulsifiers (2).

SPECIFIC GRAVITY

Specific gravity is a comparison of the weight of an oil with that of the same volume of water. The specific gravity of vegetable oils is usually about 0.910–0.920 g/ml at 25°C (77°F). As the temperature increases, the specific gravity of the oil or fat decreases. As oils/fats

Table 4.2.　**Commonly Used Food Emulsifiers**

Emulsifier	Characteristic	Processed Food
Monoglycerides/ diglycerides	Emulsification of water in oil	Margarine/cake shortening
	Antistaling or softening	Bread and rolls
	Prevention of oil separation	Peanut butter
Lecithin	Viscosity control and wetting	Chocolate
	Antispattering and antisticking	Margarine
Lactylated monoglycerides/ diglycerides	Aeration	Batters (cake)
	Gloss enhancement	Confectionery coating
Polyglycerol esters	Crystallization promoter	Sugar syrup
	Aeration	Icings and cake batters
	Emulsification	
Oxystearin	Crystallization inhibitor	Salad oil
Propylene glycol esters	Aeration	Cake batters/mixes
Sodium stearoyl lactylate (SSL)	Aeration, dough conditioner, stabilizer	Bread and rolls
Calcium stearoyl lactylate (CSL)		

are lighter than water (specific gravity less than 1,000), when mixtures of water and oils/fats separate, the oil/fat will be on top of the water.

SOLUBILITY

Oils and fats are almost completely insoluble in water. When they are held together in systems such as cake batters or butterfat in milk, these systems require the use of food emulsifiers and/or mechanical means such as homogenizers. These same oils, fats, and fatty acids are completely miscible in many organic solvents, such as hydrocarbons, ethers, esters, and so on. This is especially true at temperatures above the melting points of the fats or oils in question. With increasing temperatures, the rapidity and completeness of solubility increases.

Hexane is a very important organic solvent. It is used to extract vegetable oils from their oil seeds.

REFRACTIVE INDEX

The refractive index of fats and fatty acids is a very important characteristic because of its utility in analytical procedures. It is based on the ratio of the speed of a light wave in air as compared to its speed in the fat substance. The procedure is easy, speedy, and requires a very small sample. It is very useful for identification purposes, establishing purity, and observing the progress of reactions such as catalytic hydrogenation.

SUMMARY

This chapter covered the physical properties of oils and fats that are of most practical importance. Included were flavor, melting points, plasticity, fluidity, color, viscosity, emulsification, specific gravity, solubility, refractive index, and polymorphism. In addition to their being important in the commercial manufacture of food products, many physical methods of testing and analysis are more accurate and less time-consuming as compared with chemical analysis.

References

1. deMan, J.W. 1982. Food Microstructure *1*, 209–222 (1982).
2. Institute of Shortening Edible Oils. 1988. *Food Fats and Oils*, 6th ed. DC: Institute of Shortening and Edible Oils, p. 15.

Chapter 5

Sources of Oils and Fats

There are numerous sources of oils and fats that go into the production of food products. They can be of vegetable or animal origin. In the United States, the more important vegetable oils consumed include soybean, coconut, canola, cottonseed, and corn oils. Soybean, canola, cottonseed, and corn oils are produced in the United States, whereas coconut oil is imported primarily from tropical growing climates such as the Philippine Islands. Important meat fats include lard, edible tallow, and butterfat. Table 5.1 summarizes changes in the consumption of oils and fats from various sources from 1965 through 1991.

These data indicate the amount of food oils and fats available for consumption in the United States, sometimes referred to as "disappearance" data. Such data overestimates true consumption, because they do not consider the amounts wasted or deliberately discarded.

The greatest significance is the continued trend in the direction of products prepared from vegetable oils and away from those prepared from animal fats. Butter (80% fat) declined from a per capita availability of 6.5 lbs in 1965 to 4.2 lbs in 1991. The availability of lard for direct use declined in this period from 6.3 to 1.7 lbs.

Concurrently for this period, shortenings, which are produced primarily from vegetable oils, increased from 14.2 to 22.1 lbs. Similarly, salad and cooking oils increased from 12.5 to 25.2 lbs.

Table 5.1.   Food Oils and Fats. Per Capita Availability for
Consumption (in pounds per person)

	1965	1970	1975	1980	1985	1990	1991
Butter	6.5	5.4	4.7	4.5	4.9	4.4	4.2
Margarine	9.9	10.8	11.0	11.3	10.8	10.9	10.6
Lard[a]	6.3	4.6	3.2	2.6	1.8	2.2	1.7
Edible tallow[a]	NA	NA	NA	1.1	1.9	0.8	1.4
Shortening	14.2	17.3	17.0	18.2	22.9	22.2	22.1
Salad/cooking oil	12.5	15.4	17.9	21.2	23.5	24.2	25.2
Other oils/fats	1.6	2.3	2.0	1.5	1.6	1.2	1.3
Total Prod. Wt.[b]	51.0	55.8	55.8	60.3	67.2	65.8	66.5
Total Fat Content[c]							
Animal	NA	14.1	10.8	12.3	13.3	10.2	9.6
Vegetable	NA	38.5	41.9	44.8	50.9	52.5	54.0
Total[b]	NA	52.6	52.6	57.2	64.3	62.7	63.6

[a]Direct use excludes use in margarine, shortening and nonfood products.
[b]Computed from unrounded data.
[c]Fat content of butter and margarine is 80% of product weight.
Source: Adapted from the USDA 1982 Agriculture Outlook Oil Crops, ERS USDA
Washington D.C.

Further USDA Economic Research data indicate that in 1940 about
two-thirds of the visible fat available was from animal origin and
one-third from vegetable oils. Today, vegetable oils contribute about
85% of the visible oil or fat available for consumption.

MEAT FATS

Butterfat

As obtained from cows' milk, butter is a mixture of butterfat, water,
and salt. Creamery butter is commonly colored artificially to a consistent yellow hue. Butter made on the farm is usually not colored
and, therefore, will vary in depth of yellow color with the season
of the year and the feed of the cows. Both table-grade butter and
butter earmarked for bakery and industrial food use contain a minimum of 80% butterfat by law. The butterfat serves as a plastic matrix enclosing the aqueous phase and the solids other than butterfat.
Butter is an important source of vitamin A and, to a lesser extent,
of vitamin D. Butterfat is comprised of 29–32% monounsaturated,

2–4% polyunsaturated, and 62–69% saturated fatty acids. Butter is further characterized by the fact that over one-third of these saturated fatty acids are of relatively short chain length (14 carbons or less).

Butterfat is both desirable and expensive as a raw material. Butter's distinct flavor and yellow color are important factors in its popularity. It has a high image rating. Butterfat is also a part of many other dairy products such as milk, cheese, ice cream, coffee cream, and whipping cream.

In recent years, the use of butterfat as a table spread has been decreasing in the United States. Use of margarine has increased by a comparable quantity because of lower price, improved and uniform quality, and health factors. Butterfat, in the meantime, has been used to a greater extent in more profitable end uses, such as cheese and ice cream.

Obviously, government support and subsidy programs have provided incentive to maintain high butterfat production levels.

Lard

Lard is the fat rendered from the fatty tissues of hogs. Lard production, therefore, depends on the number of hogs slaughtered each year and the size and composition of hogs. In recent years, the demand for somewhat leaner meat cuts from the hog has tended to lessen the supply of lard. Consumption in the United States continues to decrease; therefore, about 12–15% of the lard produced in the United States is exported to other countries.

Tallow

Edible tallow is obtained primarily from beef cattle. At room temperature, it is harder and firmer than lard. Tallow does not blend readily with ingredients used in making baked food products. Therefore, tallow has limited use in baking. Some fast-food chains have had a preference for the use of certain beef tallows for deep-frying potatoes. This is due to the flavor imparted to the potatoes as well as the generally lower price per pound as compared with vegetable shortenings and oils. However, in recent years even the very large fast-food chains have been switching to vegetable shortenings or vegetable oils for health reasons. In addition, in most frying

and baking usages today, the blander flavor of the vegetable shortenings and oils is preferred over tallow. As in the case of lard, about
12–15% of edible tallow produced in the United States is earmarked
for the export market.

VEGETABLE OILS

Canola

Canola is the oil obtained from a relatively new variety of rapeseed
plant. It comes from the same type of plant as does rapeseed oil.
The original rapeseed oil was never much of a factor for edible use
in the United States. For one reason, the plant did not grow well
in our relatively warm summer climates. As it did better in cooler
climates, it was a more important edible oil in countries such as
Canada, Russia, and Finland. Another limiting factor in the original
rapeseed oil was the high level of erucic acid, ranging from 22 to
45% (1). Some research studies indicated that when substantial
amounts were fed to experimental animals, it caused some adverse
changes in the heart tissue of some of these animals. This will be
covered in greater detail in Chapter 13.

Some changes were necessary in order to increase the usage of
rapeseed oil in the United States. First, Canadian botanists developed a plant which breeded out erucic acid, resulting in canola oil.
Then researchers in the United States developed plant varieties that
perform well in the warmer climates.

In the past, U.S. farmers have been unwilling to sacrifice government subsidies by taking a risk on an uncertain crop such as canola.
This factor appears to be undergoing a change. Future farm bills are
expected to further encourage domestic canola production.

Seed manufacturers, farmers, and food processors are now working together to ensure a more consistent domestic canola supply.
Food companies have found that some U.S. consumers are willing
to pay a premium price for canola, and they are working hard to
expand the market.

The key to the acceptance of canola is its low level of saturated
fatty acids (about 6%) (2). Because of the health issue, it is possible
that canola could become the second most important source of vegetable oil early in the 21st century.

Coconut

Coconut oil is obtained from copra, which is dried coconut meat. It is classified as a fat because it is solid at room temperature, but it does become a liquid oil above 78°F (25.6°C). This fat is characterized by a high percentage of lauric acid (a 12-carbon-chain saturated fatty acid) and other fairly short-chain fatty acids. Furthermore, unlike almost all other oils and fats, coconut oil has a rather sharp melting point of about 76–78°F (24.4–25.6°C), because of a high content of low-molecular-weight fatty acids of comparable chain lengths. Other fats tend to exhibit a gradual softening with an increase in temperature (3). Because of its high level of saturated fatty acids, coconut oil is quite resistant to oxidative changes under normal storage conditions, making it a good product for cracker spraying and nut frying.

In commercial frying operations, care must be taken to make sure that coconut oil is not mixed with other frying shortenings and oils. Owing to the much lower molecular weight of coconut oil, it has a tendency to cause a foaming problem. Mixing fats and oils with considerable differences in molecular weights tends to increase the possibility of foaming. This is somewhat similar to what occurs in excessive polymerization of frying fats, in which there is a mixture of low- and high-molecular-weight fats.

Because of its sharp melting point, coconut oil is used in confections and cookie fillings. Here the sharp melting point below body temperature contributes to a "cooling" effect in the mouth.

Due to the very high level of saturated fatty acids (about 92%), the image of coconut oil has taken quite a beating over the past few years. Crude coconut oil or copra is imported into the United States, primarily from the Philippine Islands.

Corn

Although corn is one of the principal crops in the United States, only a small portion of it is used for obtaining corn oil. Most of the corn oil produced is a by-product of cornstarch production.

The primary use of this oil is in corn oil margarines, which are made and sold as premium margarine products. A high level of polyunsaturated fatty acids (about 55%) is one of the major attractions. About 20–30% of U.S. production is exported.

Cottonseed

This oil is obtained from the seeds of the cotton plant. The oil is a by-product and is dependent on the use of cotton in textiles.

Crude cottonseed oil has a strong flavor and odor and a dark reddish-brown color. It can be processed to have a bland flavor and a relatively clear color (4). However, cottonseed oil cannot be processed to have as light a color as soybean oil without destroying much of its natural antioxidant property.

Cottonseed oil is used in the preparation of some shortenings (usually a very minor portion), in some margarine oils, as a salad oil, and for deep-frying some snack items. Close to 50% of the U.S. production of this oil is exported to other countries.

Olive

Olive oil is a very important cooking and salad oil in countries bordering on the Mediterranean Sea, but it is used in relatively small quantities in the United States. It has a preference limited primarily to those who have migrated to this country from the Mediterranean area.

Virgin olive oil is oil that has not been deodorized to remove natural olive oil flavor elements, which are considered to be desirable by those using this oil.

Olive oil contains about 71% oleic acid (octadecenoic/monounsaturated) and, thus, stakes out its own health or nutritional claim. See chapter 13.

Palm and Palm Kernel

Palm oil is a fruit oil (like olive), as compared with seed oils. It is one of the more important fruit oils in world trade and its use grew at a rapid rate for a period of time. Palm oil is a semisolid oil extracted from the pulpy portion of the fruit. The fruit grows in large bunches or clusters, with a typical bunch weighing about 40 lbs. Each individual fruit is about the size of a date, and the fruit mesocarp generally has an orangelike color when properly ripened. The inside kernel contains the palm kernel meal and the palm kernel oil. Palm kernel oil is different from palm oil and more like coconut oil in performance properties.

New strains or hybrids are under cultivation, especially in Malaysia, that should improve the overall yield of palm oil and the ratio of palm oil produced to palm kernel oil produced. In addition, with good cultivation techniques, palm trees yield more oil per acre than any other oil-bearing plant—typically from 1 to 1.5 tons/acre, compared with soybean yields of 300 lb/acre. In addition, once the trees are planted, little upkeep is required over their 30–40 years of productive life.

Palm trees and palm oil production are native to tropical West Africa: However, recent innovations and growth have occurred in Indonesia, in addition to Malaysia. Palm trees require a tropical climate and about 100 in. of rain per year. At one time it was thought that palm oil would provide serious competition to soybean oil in the United States. As a result, various members of Congress and special interest groups, lobbyists, and so on, became very nervous. There were hues and cries to limit imports, set up duties, and increase subsidies of our crops; however, the panic was apparently for naught. One reason was that the consumption of fats and oils in Europe, India, Japan, and other Oriental countries increased by such a rapid rate that these countries imported Malaysian palm oil as fast as it could be processed. The second reason is the fact that palm contains about 50% saturated fatty acids (5).

Palm oil is used as part of the fat/oil portion of shortening and margarine. It is also an important cooking oil in Europe and the Orient.

Peanut

The major peanut crop is located in the southeastern section of the United States. The crude oil is pale yellow and has the characteristic odor and slight flavor of peanuts. It is a relatively unimportant edible oil in this country. It also tends to be quite an expensive oil in relation to its potential use—primarily for deep frying and as a cooking oil. About 40–50% of the modest production of peanut oil in the United States goes to the export market.

Safflower

Safflower oil has been known since ancient times, but only in the past 30 years has it assumed any degree of commercial importance.

Since 1960, the United States acreage has grown slightly in Nebraska, Colorado, and California, but now it appears to have leveled off at a low production level. Its slight growth can be attributed to its high linoleic (polyunsaturated) acid content—about 75–80%, which is the highest of the commercially available oils. Its use has been very limited because of short supply, high cost, and lack of flavor stability.

Soybean

This oil is obtained from the seeds of the soybean plant. It has grown from an unimportant oil before World War II to the most important vegetable oil produced in the United States (6). It represents over 65% of the vegetable oil usage in this country and constitutes about 55% of the worldwide use of vegetable oils.

The major uses are in the manufacture of vegetable shortenings for baking and frying, fluid vegetable shortenings, margarines, and salad and cooking oils.

Soybeans can be grown easily under a wide variety of soil and climatic conditions. Production of bumper crops was accelerated by the U.S. government support of the domestic price of soybeans during the period of rapid growth. This program resulted in rapid technological advances in the growing of soybeans and in the processing of soybean oil. This, in turn, resulted in rapid expansion of both domestic usage and export trade. From 10% to 25% of the United States production is exported to other countries, including the Soviet Union.

Sunflower

The sunflower plant is grown for oil production in relatively large quantities in the former Soviet Union, Argentina, and Hungary. In recent years, sunflower growth in the United States has increased at a modest pace. It is a relatively easy crop to grow, but the resulting oil has different properties, depending on the climatic conditions. So far, the most desirable oil for use in salad oil has come from the colder climates of the Dakotas and Minnesota.

The growing popularity of sunflower oil appears to be primarily related to the following:

1. It has good flavor stability without a need for hydrogenation. For example, sunflower oil has better flavor stability than safflower oil even though both contain about 90% unsaturated fatty acids. This is due to the fact that sunflower oil contains lesser amounts linoleic acid (55–60% vs. 75–80%) and about double the oleic acid (about 30% vs. about 15%).

2. It has a high percentage of unsaturated fatty acids, including 55–60% polyunsaturates as linoleic and about 30% monounsaturates as oleic acid. It has been implied that eliminating hydrogenation makes it a more "natural" oil.

On a worldwide basis, the USDA continues to predict record consumption levels for soybean, palm, sunflower, and olive oil during the 1990s. In addition, this USDA report indicates that the long-term outlook for vegetable oil consumption holds a great potential, because "most of the world is on a low-fat diet" (7).

SUMMARY

Although oils and fats are essential components of almost all forms of plant and animal life, there are relatively few that produce oil and fat in sufficient quantity, or in a sufficiently available form, to cause them to be an important article of commerce. Those that qualify to a greater or lesser extent are reviewed in this chapter.

At the present time, soybean oil is by far the most important oil/ fat produced and used in the United States. In some other parts of the world, the most important products are as follows:

Hungary: About 80% of the oilseed acreage is in sunflower seed (8)

Canada: Mainly canola oil

Mexico: Meat fats

Europe (EEC except Spain): Rapeseed, sunflower, soybean, and peanut oil

Spain: Olive oil

Philippines: Coconut oil

Pacific Rim: Palm, palm kernel, and Indian soybean oil

References

1. Dotsun, K. 1991. Inform 2(7), 611.
2. Dotsun, K. 1991. Inform 2(7), 611.

3. Swern, D. 1979. *Bailey's Industrial Oil and Fat Products, Vol. 1,* 4th ed. New York: Wiley-Interscience, p. 312.

4. Swern, D. 1979. *Bailey's Industrial Oil and Fat Products, Vol. 1,* 4th ed. New York: Wiley-Interscience, pp. 352–353.

5. Van Duyvenbode, H. and Sutter, J. 1993. Oils & Fats Int., Issue One.

6. Gupta, M. 1993. Inform 4(11), 1267.

7. Inform, 1992 3(12), 1314–15.

8. Inform, 1992 3(11), 1180.

General References

Shahidi, F. 1990. *Canola and Rapeseed Production, Chemistry,* Nutrition, and Processing Technology. New York: Van Nostrand Reinhold.

Chapter 6

Processing Technology

This chapter treats the major steps in the manufacture of short-enings, oils, and related finished food fats. Some of these steps will vary, depending on the desired properties of the end product. A basic knowledge of the manufacture of these products will provide a good background for understanding how they differ and how they can be used to best advantage in food preparation and food processing.

EXTRACTING THE OIL/FAT

Food oils and fats originate from oilseed and animal sources. Animal fats (lard and tallow) are rendered from animal tissues during the course of slaughtering and processing. Rendering may be done with either dry heat or steam (1). Rendering and processing of meat fats is conducted in plants inspected by the U.S. Department of Agriculture (USDA). Lard and tallow generally contain water and protein that must be removed and relatively high levels of free fatty acids that must be greatly reduced.

Vegetable oils are extracted from seed fruits or nuts. In the United States today, this is accomplished primarily through solvent extraction. In the past, it was done primarily by physical means, such as the application of heavy pressure to the seed fruit or nut. Palm oil, which is imported primarily from Malaysia, is still separated by physical means such as hydraulic presses.

Canola oil (rapeseed) as processed in Canada has been extracted by a combination operation including both pressing and solvent extraction. Canola seeds are flaked, cooked, rolled, semipressed, and then solvent extracted.

Solvent extraction as employed in the United States is the most efficient means of deriving oil from the soybean or cottonseed. Generally, 9–12% more oil with fewer impurities can be extracted from soybeans or cottonseeds by solvent extraction than by mechanical pressing. Further, a minimum of heat is involved in solvent extraction, so the oil produced is of better quality. This process is energy conserving as compared with mechanical extraction (2).

Hexane is the most widely used solvent. The oil fraction of the soybean or cottonseed is soluble in hexane during extraction, and the hexane is distilled off and reused. Because of its high volatility, little or no hexane residue remains in the finished oil after processing.

The oils or fats obtained from the extraction of oilseeds and from rendering animal fats are called "crude" oils and fats. They contain varying quantities of naturally occurring materials other than triglycerides. Included are free fatty acids, small amounts of protein, phospholipids, phosphatides, waxes, resins, color pigments, and flavoring substances that are undesirable elements in the final food products. Although present in relatively small amounts, these materials contribute undesirable colors, flavors, odors, instability, and foaming and smoking during deep-frying. Therefore, all of these materials must either be eliminated or materially reduced during further processing stages.

Generally, crude oils are degummed to remove phospholipids/ phosphatides at the oil mill or extraction plant. These phospholipids/phosphatides are hydrated with small amounts of water to make them insoluble in the oil. Sometimes citric acid and/or phosphoric acid is agitated with the water to improve speed and efficiency and to chelate metals. The phospholipids/phosphatides are then separated from the crude oil by continuous centrifuging. In the case of soybean oil, this also creates a very marketable by-product, soya lecithin (3). Cottonseed lecithin is generally too dark in color to be of commercial value.

OIL PURCHASE

Before purchasing, samples of the crude oils are generally completely tested by the refining company. For example, a miniature refining test is run to determine the expected level of free fatty acids, color, flavor, and other important characteristics in the oils' final forms. These tests determine their ultimate quality as finished brands. In addition, tests can determine for which type of finished oil or fat products they are suitable.

Generally speaking, it is also wise for companies to inspect certain suppliers, such as their packaging manufacturers and those furnishing additives. Companies should monitor cleanup and the inspection of conveyances of bulk shipments.

Oils and fats as raw materials for finished products are relatively low-risk materials from a regulatory viewpoint. Low risk in this context refers to the clauses in the Food, Drug, and Cosmetic (FD&C) Act that relate to potential contamination. The key reasons for the low risk associated with these materials are as follows:

1. The comparative unattractiveness of these materials as sources of food to visible pests (mostly insects and rodents).
2. The fact that the processing techniques for these materials (chemical treatment, filtration, use of elevated temperatures, etc.) all tend to destroy or remove contaminants.
3. The use of essentially enclosed processing systems for these materials protects them against contamination from the plant environment.
4. The fact that current analytical technology makes any adulteration of these materials easy to detect.

The section of the FD&C Act dealing with the presence of poisonous or deleterious material is the one most likely to be of concern, as opposed to the section dealing with the presence of filth or with the potential for contamination. This is not a likely possibility, as it has been years since any episodes involving crude oils and fats have been reported in the FDA's Weekly Enforcement Report. Some small risk remains, however, and is likely to involve such materials as pesticide residues or lubricants.

Finished oil and fat products are of a somewhat higher risk—particularly from the potential for contamination existing in the materials used to package them such as cans, bottles, and so on. As a result, periodic inspections of suppliers of packaging materials including shipping containers should certainly be considered.

OIL RECEIVING

Shipments of crude oils into the refiner's plant should be scheduled in order to minimize holding time and processing. It has been found

Table 6.1.   Some Processing Aids Used in Manufacturing Edible Fats and Oils

Aid	Effect	Mode of Removal
Sodium hydroxide	Refining aid	Acid neutralization
Carbon/clay (diatomaceous earth)	Bleaching aid	Filtration
Nickel	Hydrogenation catalyst	Post bleach and filtration
Sodium methoxide	Rearrangement catalyst	Water or acid neutralization, filtration, and deodorization
Phosporic acid ⎫ Citric acid ⎬	Refining acids, metal chelators	Neutralization with base, filtration, or water washing
Acetone ⎫ Hexane ⎪ Isopropanol ⎬ 2-Nitropropane ⎭	Crystallization media for fractionation of fats and oils	Solvent stripping and deodorization
Nitrogen	Oxygen replacement	Diffusion
Polyglycerol esters	Crystallization modification	Caustic refining and filtration
Sodium lauryl sulphate	Fractionation aid, wetting agent	Washing and centrifugation

Courtesy of Institute of Shortening and Edible Oils.

that maximum flavor stability and frying stability are dependent on the speed of processing. The refiner should minimize the time between the extraction of the vegetable oils and complete processing and packaging of the specific food fat. This also makes sense in terms of the economies of maintaining a minimum inventory of crude oils and oils in the intermediate processing steps. (See Table 6.1.)

REFINING

The first step in the plant refining procedure is to react the crude oil with an alkaline material to remove the free fatty acids. (See Fig. 6.1) An excessive quantity can contribute to an unsatisfactory flavor in a fat and detract from a fat's frying life. When the very weak alkali of an exact measured strength reacts with the free fatty acids, the result is soap.

$$RCOOH + NAOH \rightarrow RCOONA + H_2O$$
Free fatty acid   Alkali   Soap   Water

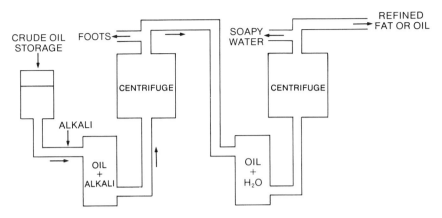

FIGURE 6.1. Refining procedure.

This soap must be removed, and this is done by passing the mixture of fat and soap through a continuous centrifuge machine that separates the fat or oil from this soap, which is sometimes referred to as "foots." The refined far or oil is water washed to remove the final traces of soap and is centrifuged again. Finally, the refined material is dried to remove the residual water.

Other very minor materials, such as phosphatides, proteins, carbohydrates, waxes, and gums, that have not been previously removed by the degumming process are also separated out during refining. Oils that are low in phosphatide content, such as palm or coconut, may be physically refined (i.e., steam stripped to remove free fatty acids) (4).

The latest continuous refining techniques have greatly contributed to the efficiency of refining, the speed of refining, and the improved quality of the refined oil.

BLEACHING

The term "bleaching" refers to the treatment that is given to remove colors and color-producing substances. This is done so that the finished fluid shortenings, plastic shortenings, or salad oils will be light in color. The prime purpose is to remove the color pigments, such as reds, yellows, and greens, that are present in the crude oil (5).

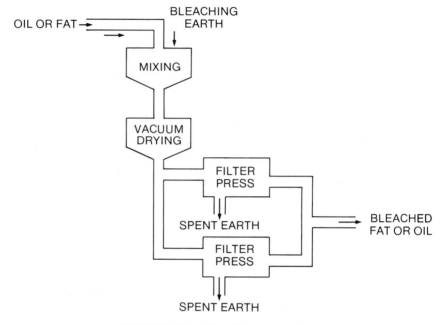

FIGURE 6.2. Bleaching procedure.

The usual method of bleaching is by adsorption of the color pigments on an adsorbent material. Fuller's earth, for example, is a natural bleaching earth consisting basically of hydrated aluminum silicate. Within recent years, natural earths or clays have been replaced to a considerable extent by acid-activated clays or earths that have considerably more bleaching power. Activated carbon is also used as a bleaching adsorbent to a limited extent. The color pigments are adsorbed onto the earth or clay, and then the earth or clay is filtered from the oil. This is shown in Fig. 6.2. Note the continuous testing of these oils throughout all steps in processing.

Continuous methods of bleaching are generally more efficient than the older batch-bleaching methods.

HYDROGENATION

This is the manufacturing process that permits the production of both plastic solid shortenings and fluid shortenings from liquid oils.

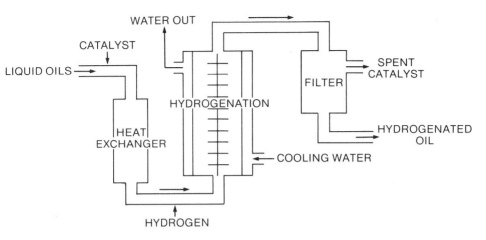

FIGURE 6.3.   Hydrogenation.

Hydrogen adds directly to the points of unsaturation in the fatty acids. The hydrogenation process (Fig. 6.3) was developed as a result of the need to (1) increase the oxidative stability of the oil or fat and (2) convert liquid oils to fluid shortening and plastic shortening by increasing solids for greater utility in certain food uses (6).

Hydrogenation is extremely important in using soybean oil, our most important source oil. Soybean oil now provides 60–65% of the total visible fat in the U.S. diet. Crude soybean oil contains about 8% linolenic acid, which is very reactive with oxygen. The linolenic acid content must be reduced to around 2% or less to have acceptable oxidative/flavor stability properties in most food operations, and hydrogenation is presently the only practical way to achieve this.

Fatty acids with two double bonds, such as linoleic, can be hydrogenated to form fatty acids with one double bond, such as oleic, or with no double bonds, such as stearic. Likewise, fatty acids with three double bonds (such as linolenic) can be hydrogenated to form fatty acids with two double bonds, one double bond, or no remaining double bonds. As the degree of hydrogenation increases, the saturation of the fat and its firmness at room temperature also increase. Obviously, its melting point increases too, as does its degree of oxidative stability under storage conditions.

Hydrogenation is accomplished in a reactor where hydrogen gas is bubbled through the oil at a suitable temperature and pressure in the presence of a catalyst. The catalyst accelerates the reaction but is not consumed in the reaction. It may enter into temporary combination with the reactants, but such combinations are unstable and are invariably broken down at the completion of the reaction to again yield the catalyst in a relatively unchanged form. Therefore, the catalyst may be used over and over again until it reaches the point that it loses its efficiency.

Through careful selection of temperature [250–450°F (121–232°C)], pressure [10–60 psi. (gauge)], and type of catalyst (usually nickel supported on an inert carrier), the oil may be selectively hydrogenated to obtain the desired characteristics. The oil may be selectively hydrogenated to also minimize the development of trans acids or to maximize the conversion of linolenic acid to linoleic acid when either is desirable. The degree of hydrogenation is predetermined, depending on the type of intermediate or final product desired. The hydrogenation process is easily controlled by refractive index measurements of the oil and it can be stopped at any desired point.

If an oil is completely hydrogenated, it will be a very hard, brittle solid at room temperature. Plastic shortenings are only partially hydrogenated to obtain the proper balance between flavor stability and a workable consistency over wide usage temperatures. It is customary to blend materials that have received different degrees of hydrogenation (intermediate "stocks") in order to obtain the desired properties in the finished oil or shortening. For example, salad oil would be made up of only one stock, fluid shortening generally two stocks, and plastic shortening could be made from two, but usually three or more intermediate hydrogenated stocks.

Fluid shortenings are made from soybean oil stocks of two different degrees of hydrogenation to obtain the proper balance among flavor stability, pourability, and frying performance. Plastic shortenings may be composed of soybean oil hydrogenated to two or three different end points, and palm oil or cottonseed oil of one or two different consistencies or degrees of hydrogenation. This wide range of stocks hydrogenated to different degrees is used to produce shortening with a wide plastic range—workable and usable over a wide temperature range.

In making fluid shortening, soybean oil is generally used to facilitate getting the finished product into the beta-crystalline phase in which it is pourable and pumpable. A 100% soybean product is beta-tending.

In making plastic shortenings, some hydrogenated palm oil or hydrogenated cottonseed oil must be included in the formula to facilitate getting the final product into the beta-prime crystalline phase. Beta-prime crystals are small and needlelike and hold the finished product in a plastic form. Palm oil and cottonseed oil are beta-prime-tending. This will be covered in greater detail in the section "Plasticizing" in this chapter.

Continuous methods of hydrogenation are employed in making one single stock over prolonged periods of time without change-overs. In plants that produce varieties of finished, hydrogenated stocks, batch-hydrogenation equipment is more efficient.

FRACTIONATION
(INCLUDING WINTERIZATION)

Fractionation is the removal of solids at selected temperatures. The most widely practiced form of fractionation is that of crystallization wherein a mixture of triglycerides is separated into two or more different melting fractions based on solubility at a given temperature. The term "dry fractionation" frequently is used to describe fractionation processes such as winterization or pressing. Winterization is a process whereby a small quantity of material is crystallized and removed from edible oils by filtration to avoid clouding of the liquid fractions at refrigeration temperatures. Originally, this processing was applied to cottonseed oil by subjecting the oil to ambient winter temperatures, hence the term "winterization." Today many oils, including cottonseed and partially hydrogenated soybean oils, are winterized. A similar process called "dewaxing" can be utilized to clarify oils containing trace amounts of clouding constituents (7).

Pressing is also a fractionation process sometimes used to separate liquid oil from solid fat. The process squeezes or "presses" the liquid oil from the solid fat by means of hydraulic pressure. This process is used commercially to produce hard butters and specialty fats from such oils as palm kernel and coconut.

Solvent fractionation is the term used to describe a process for the crystallization of a desired fraction from a mixture of triglycerides dissolved in a suitable solvent. Fractions may be selectively crystallized at different temperatures after which the fractions are separated and the solvent removed. Solvent fractionation is practiced commercially to produce hard butters, specialty oils, and some salad oils from a wide array of edible oils.

Cottonseed oil, for example, must be winterized if it is to be used in making a salad oil. The high melting point solids normally present would crystallize out and form a cloud in the oil when allowed to cool in the refrigerator. In addition, if such an oil is used in making mayonnaise, these higher melting fractions will solidify when the mayonnaise is refrigerated, which will result in breaking the emulsion and, thus, causing separation in the mayonnaise. If cottonseed oil is used only as a cooking oil, it would not require winterization.

Soybean oil is considered a natural winter oil because it does not have a tendency to crystallize at refrigeration temperatures. Quite often soybean oil is refined, bleached, and deodorized, but not hydrogenated or winterized in making a salad oil. However, if a soybean salad oil is also to be used in deep-frying, it should be lightly hydrogenated to reduce the linolenic acid content from about 8% down to about 2%. The oil undergoes very selective and partial hydrogenation and then is subjected to the winterization process. Dimethyl polysiloxane may be added in a few parts per million to control foaming while frying.

DEODORIZING

Deodorizing is the process that makes possible the production of a bland, neutral flavor so that the finished shortening or other fat food product will not impart any flavor of its own to the final fried, baked, or other food product.

Each product to be deodorized has been formulated from nonhydrogenated oils and/or hydrogenated stocks that have received their proper degree of hydrogenation.

Deodorizing is accomplished in a "still" or steam distillation chamber that literally boils or strips away objectionable odors and flavors present in the refined, bleached, and hydrogenated oil. This is feasible because of the differences in volatility between the triglycerides and the substances that give flavors and odors to fats (See Fig. 6–4.) Refined, bleached, and hydrogenated oils contain minor quantities (less than 0.1%) of materials that could impart undesirable flavors or odors to these oils. These materials are easily removed in the distillation process, which is carried out under low pressure and high temperature (about 400–550°F; 204–288°C). Deodorizing is the final step in processing and is done just before plasticizing and packaging (8).

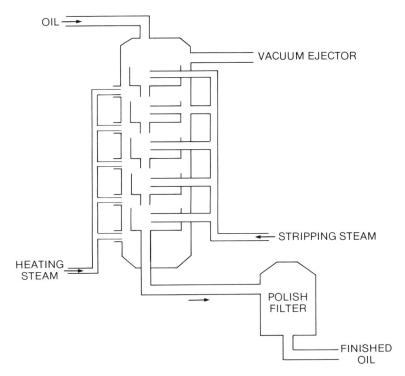

FIGURE 6.4. Deodorizing

Continuous, semicontinuous, and batch deodorizing equipment are employed. The selection is usually based on the number of changeovers in a specific plant.

PLASTICIZING

In making plastic shortenings, it is important to develop a fine needlelike crystal structure (beta-prime form) in order to produce a shortening that is smooth in appearance and firm in consistency. This is accomplished through the step of chilling the properly formulated fat very rapidly in scraped-wall heat exchangers, a piece of equipment resembling an ice cream freezer. During this rapid cooling, it is customary, along with vigorous agitation, to whip in air (or nitrogen) in the amount of 10–15% of the shortening's volume.

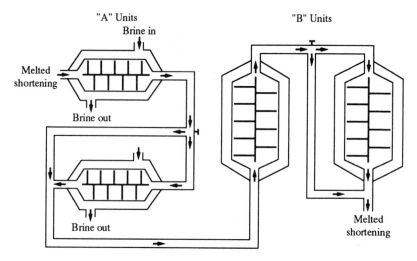

FIGURE 6.5.   Plasticizing

The shortening is usually packaged in 1- and 3-lb retail packages, or 50-lb cartons with a polyethylene bag for the baker or food processor. After packaging, the shortenings are "tempered" for 24–72 hours at about 85°F (29.4°C) to further ensure proper crystal growth.

A typical plasticizing operation is shown in Fig. 6.5. This shows two "A" units and two "B" units that are used to produce about 10,000 lbs of shortening per hour. The completely formulated shortening, which has been stored above its complete melting point (130–140°F; 55–60°C), is allowed to flow by gravity into another tank equipped with a level-controlling float from which it is pumped through a precooler and cooled to 110–115°F (43–46°C). Then the product is pumped through the "A" unit or units. These "A" units are sometimes called freezers, chillers, or Votators. Here the product is supercooled in a matter of about 18 seconds to 65–70°F (18–21°C) with vigorous agitation. Then this supercooled oil is passed through a "B" unit or units (worker units) in about 3 min under moderate agitation. During residence in the "B" units the latent heat of crystallization released causes the temperature to rise 10–15°F (5.6 to 8.3°C). In addition to causing the oil to solidify into small needle-like crystals, the agitation in the "B" units is needed to distribute the heat of crystallization uniformly throughout the product.

FIGURE 6.6.   Left: beta-crystals; right: beta-prime-crystals.

The basic difference between plastic shortening and fluid beta-shortening lies in the polymorphic forms of the fat crystals in the two products. In plastic fats, the crystalline structure is in the beta-prime polymorphic form, with long, needlelike crystals. These needlelike crystals intermesh to form the physical structure of the plastic fat.

In the fluid product, the crystals are in the beta-polymorphic form, with platelike crystals that are not subject to intermeshing; thus the product has no rigid structure and remains fluid. To form these beta-crystals, the hot, melted shortening blend is chilled rapidly in a freezer, as with the plastic fats. This chilled product is then subjected to a tempering step with constant agitation, during which the crystals are converted to the beta-phase. This takes several hours to complete. Fig. 6.6 is magnified to show the difference between the beta- and beta-prime-crystal structure. The product is then ready for packaging. Chapter 4, in a section on polymorphism, covers this process in more detail. For a further description and understanding of the Votator "A" and "B" unit, please refer to Fig. 6.7 (a cutaway

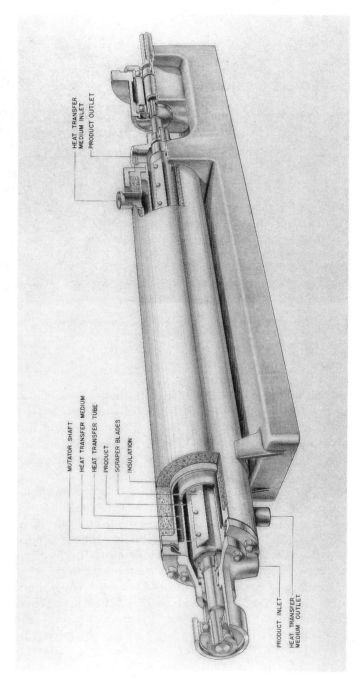

HEAT TRANSFER
MEDIUM INLET
PRODUCT OUTLET

MUTATOR SHAFT
HEAT TRANSFER MEDIUM
HEAT TRANSFER TUBE
PRODUCT
SCRAPER BLADES
INSULATION

PRODUCT INLET
HEAT TRANSFER
MEDIUM OUTLET

FIGURE 6.7.   Cutaway view of a Votator scraped-surface heat exchanger ("A" Unit). (Courtesy of Cherry-Burrell.)

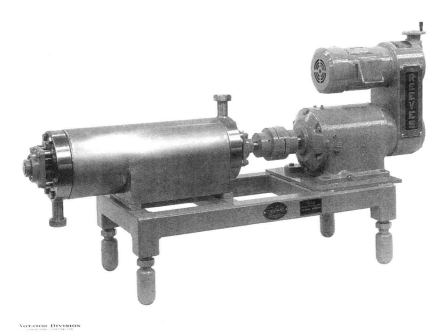

VOTATOR DIVISION

FIGURE 6.8.   Votator "B" Unit for shortening and margarine processing. (Courtesy of Cherry-Burrell.)

view of the "A" unit), Fig. 6.8 (a Votator "B" unit) and Fig. 6.9 (a cross section of a Votator unit.)

PACKAGING

Plastic shortening is generally packed in 1- and 3-lb retail packages or in 50-lb cubes using a polyethylene inner liner.

Fluid shortening and salad oil are packed in 35-lb plastic jugs inside a protective carton, and 1-gal plastic jugs packed two to a carton. There are still some fluid shortening products packed in 5-qt cans or 5-qt foil-fiber containers, but these packages are decreasing in usage. Another package that was popular in the past, the 55-gal drum, is rarely used today. The larger food processor often takes deliveries in tank trucks or rail tank cars with capacities of 20,000 to 150,000 lbs.

VOTATOR DIVISION
C-ICHETRON CORPORATION

FIGURE 6.9.   Cross section of Votator unit. (Courtesy of Cherry-Burrell.)

SUMMARY

It must be emphasized that manufacturing can significantly affect the utilization, consistency, flavor stability, and frylife of oils and fats. Each step must be taken to maximize protection during processing and storage. The quality of the crude oil, proper processing, speed of processing, and the correct type, amount, and addition of methyl silicone result in maximizing the eventual fry life of oils and fats used for frying.

References

1. Swern, D. 1982. *Bailey's Industrial Oil and Fat Products, Vol. 2*, 4th ed. New York: Wiley-Interscience, p. 178 and 188–191.
2. Swern, D. 1982. *Bailey's Industrial Oil and Fat Products, Vol. 2*, 4th ed. New York: Wiley-Interscience, pp. 215–227.
3. Swern, D. 1979. *Bailey's Industrial Oil and Fat Products, Vol. 3*, 3rd ed. New York: Wiley-Interscience, p. 732.
4. Institute of Shortening Edible Oils. 1988. *Food Fats and Oils*, 6th ed. Institute of Shortening and Edible Oils, Washington, DC: p. 13.
5. Institute Shortening Edible Oils. 1988. *Food Fats and Oils*, 6th ed. Institute of Shortening and Edible Oils, Washington, DC: p. 13.
6. Institute of Shortening Edible Oils. 1988. *Food Fats and Oils*, 6th ed. Institute of Shortening and Edible Oils, Washington, DC: pp. 13–14.
7. Institute of Shortening Edible Oils. 1988. *Food Fats and Oils*, 6th ed. Institute of Shortening and Edible Oils, Washington DC: p. 13.
8. Applewhite, T. 1985. *Bailey's Industrial Oil and Fat Products, Vol. 3*, 4th ed. New York: Wiley-Interscience, p. 128.

Chapter 7

Deep Fat Frying

INTRODUCTION

With this chapter we start the coverage of how these oils and fats are used in edible food products. This chapter will be devoted to deep fat frying with the emphasis on the basic principles involved in this method of cooking. Chapters 8 through 12 will include such topics as griddling, pan frying, salad oil, cooking oil, baking, doughnut technology, and some large commercial uses for food fats and oils.

Food oils and fats enter into three major areas of food preparation: (1) food processing plants, (2) foodservice kitchens, and (3) the home. For some uses there are similarities in required product and performance characteristics in these three areas. However, oil and fat package sizes are obviously different, with the smallest packages being designed for home use and, at the other extreme, the large food processor receiving as much as 150,000-1b rail car shipments of oils and fats.

Food processing plants use oils and fats to produce foods for both the home and the foodservice industry. Some are ready to serve and some require further cooking/processing. Some are shelf stable and some are frozen. Some are completely cooked and some are partially cooked. Some merely require reheating. Prepared mixes, for example, require reconstitution, mixing, and baking.

The foodservice industry uses oils and fats directly in food preparation such as deep-frying, dressings, cooking, and baking. They also use intermediate products where the fat is already in the food. Examples of the latter include cake/muffin mixes, precooked/partially cooked foods requiring reheating, frying, baking, and so on.

Chapter 7 will be devoted to the basic principles of deep-frying and will include some quantity recipes and production formulas where appropriate for illustrative purposes.

THE IMPORTANCE OF DEEP FAT FRYING

Deep fat frying is a very important method of cooking because it is fast, convenient, and deep-fried foods are generally liked for flavor and texture. In commercial food processing the deep-frying process is employed for a wide range of foods which are ultimately sold to supermarket chains (plus "quick stop" chains and independent grocery outlets) and to foodservice operators. Examples include the many types of french fried potatoes, fried and frozen seafoods, poultry, egg rolls, and meat patties. Potato chips, corn chips, doughnuts, and a multitude of other snack items are processed and distributed in this manner. In fast-food operations, entrees, potatoes, and many side dishes are fried and served almost to order. Most table service restaurants offer two or three fried entree items as well as potatoes, onion rings, and some appetizers that are deep-fried.

Having the deep-fried foods prepared by food processors and foodservice operations eliminates the difficulties involved in preparing them at home. This includes handling hot frying shortening and possible problems with grease splashing, burns, inadequate ventilation, and other safety considerations. In addition, the breading and battering of foods for frying can be messy and can contribute to cleanup problems. In addition, more women are working outside the home and have less time for lengthy meal preparation.

Deep-frying is economical in that all the heat is concentrated in a small unit and there is little waste of gas or electricity. Deep-frying is a very fast method of cooking.

THE GENERAL MECHANISM OF DEEP-FRYING

Deep fat frying is a process of cooking involving the direct transfer of heat from hot fat to cold food. This is accomplished in food processing operations with a wide range of fryers with frying oil capacities of from 100 lbs to several hundred pounds. The fryers most commonly used in restaurants hold from 15 to 60 lbs of shortening. The shortening is heated directly by either electric heating elements or a gas flame inside tubes that go either through or around the kettle. The heat is regulated by thermostats, which send gas or electric heat to the frying fat in order to maintain the desired temperature.

Because of the direct application of heat from the frying fat to the food, the cooking process is rapid. When cold food is dropped into hot shortening, several things occur:

1. Heat continues to transfer even after the food is cooked and is removed from the kettle. In large frying operations the food is often conducted through the fryer on conveyor belts.

2. The temperature of the fat decreases and the thermostat is signaled to provide additional heat energy to return the fat to its temperature setting.

3. Moisture from the food starts to form steam, which is evaporated with a bubbling action that gradually subsides as the food becomes cooked.

4. A desirable browning or caramelization of the surface of the food takes place.

5. The food absorbs fat during the cooking process. Usually from 4% to 30% of the finished weight of the cooked fried food is absorbed fat. For most foods, the greatest proportion of this absorbed fat tends to accumulate near the surface of fried food. This fat adds a desirable texture to the food and provides a satisfying eating quality. The amount of fat absorbed is affected by the time the food takes to cook, the surface area of the food, the finished moisture content of the product, and the nature of the food (e.g., the types of breading and battering materials used).

6. Changes occur in the frying fat as it undergoes use.

FAT CHANGES AND REACTIONS DURING FRYING

The fats used for deep-frying gradually undergo certain chemical changes during use. The most important changes are (1) color formation, (2) oxidation, (3) polymerization, and (4) hydrolysis. In addition to these chemical changes, other physical changes such as odor and flavor development are noted.

Color Formation

All foods that are fried contribute substances (e.g., sugars, starches, proteins, phosphates, sulfur compounds, and trace metals) that collect in the fat during the frying process. These extracted materials then brown by themselves and/or react with the fat itself and cause the fat to darken.

As the fat gets darker with use, the foods fried in it darken at a more rapid rate, eventually reaching the point where the food may be too dark in color or not be completely cooked. The fried food may also take on an undesirable appearance—gray, dull, uneven color, and so on. The rate of color formation in the fat is different for various foods. Potatoes form little color in the frying fat. Chicken accelerates color darkening at a far greater rate than do potatoes, because protein-containing components cause darkening at a more rapid rate than starch. In addition, the composition of the batter and breading materials used with chicken accelerate this change. In the case of scallops and many other seafoods, it is the protein components, batter/breading composition plus phosphates and sulfur compounds, that promote an even greater degree of color darkening.

The type of breading used can have an important influence on the rate of color formation in the fat and the time it takes for the food to brown. Breadings high in "reducing sugars" such as honey, glucose (dextrose), and corn syrup solids form color much more rapidly than breadings that do not contain such substances. Bread crumbs darken fat at a more rapid rate than cracker crumbs.

The rate of color darkening or any other change in the characteristics of the frying fat depends to a considerable extent on the rate of turnover of the fat in the kettle. When any food is deep fried, some of the oil or fat is absorbed by the food. This absorbed fat must then be replaced by fresh fat. The more rapid the rate of replacement, the lower the level of color darkening. In some operations where this rate of fresh replacement is rapid enough, the oil or fat never has to be discarded, and an equilibrium color level for that specific system is reached. In food processor frying operations such as for potato chips, doughnuts, and frozen poultry and meats, the no-discard situation is generally achieved. On the other hand, in foodservice frying, this goal is not easily achieved.

Oxidation

Oxygen from the air reacts with the fat in the fryer. Some of the reaction products are removed from the fryer by the steam that evolves during the frying of the food, but other reaction products remain in the fat and can accelerate the further oxidation of the fat. At room temperature (70–80°F; 21.1–26.7°C), oxidation is generally a relatively slow process. However, at frying temperatures, oxidation can proceed quite rapidly. The higher the temperature, the more rapid the rate of oxidation. Therefore, it is important that the temperature of the frying fat not be any higher than required to do the cooking job. Most foods can be fried at 350°F (177°C) in foodservice operations. In food processing plants, frying temperatures may be well over 400°F (204.4°C). This is because many of these foods are conveyed through the frying oil in as little as 1–2 min.

Other factors besides temperature that affect the rate of oxidation include

1. The rate at which fat is absorbed by the food from the system and replaced by fresh fat
2. The amount of surface area of the fat that is exposed to oxygen
3. The presence of metals such as copper and brass, which accelerate oxidation (pro-oxidants)
4. The presence of high-temperature antioxidants such as methyl silicone, which retards oxidation (1)
5. The quality of the frying fat

Therefore, in order to hold the level of oxidation in the frying fat to a minimum, it is important to use a good-quality frying fat, keep the temperature of the fat as low as possible, follow frying rules that permit the maximum rate of turnover of fat, and avoid copper contamination. The regular removal of food particles from the frying fat is also important.

Polymerization

Excessive oxidation is often accompanied by polymerization. As oils and fats undergo heating in the deep-frying process, various de-

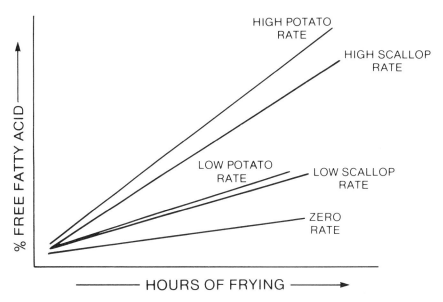

FIGURE 7.1. Foam development during deep fat frying.

composition products are formed. Some of these products are essentially volatile at frying temperatures and have relatively little responsibility in developing polymers. These volatiles include peroxides, monoglycerides and diglycerides, aldehydes, ketones, and carboxylic acids. The nonvolatile decomposition products include polar compounds, monomers (cyclic and noncyclic), dimers, trimers, and other high-molecular-weight compounds; these reactions can result in the formation of very large molecules (2).

These very large molecules, or polymers, may result in gumming. When gumming occurs, it generally appears along the sides of the fryer, on frying baskets, or on conveyor belts, where the surface of the oil or fat and the metal come in contact with oxygen from the air.

Polymerization can also result in foaming. With the development of more and more very large-molecular-weight polymers, the frying oil will contain fatty acids of considerably different chain lengths. It is this difference in chain lengths that results in frying oils/fats foaming. Excessive proteins in the frying fat may also contribute to foaming (3), but this is a minor factor.

Foaming is the formation of small bubbles that creep slowly up the sides of the fryer. If the fat foams excessively, it should be discarded even if food flavor is satisfactory because foaming fat can pose a very serious safety and fire hazard. Foaming should not be confused with the bubbling that results from the violent evaporation of moisture from fried foods. Evaporation bubbles are fairly large and tend to stop after much of the moisture has left the food.

Figure 7.1 illustrates the effect of the amount of food on the foaming of a frying fat. Notice that the zero rate (no food frying—heating only) shows a rather high rate of oxidation, polymerization, and foaming. When little or no food is fried, the frying fat does not get the "protection" of steam generated by the food or the fat turnover created by fat absorption. The "low" rate of frying scallops and potatoes shows an increase in the rate of foaming. This is due to the introduction of substances from the food that accelerate the oxidation, without enough food being fried (or enough fat absorbed and replaced by fresh) to compensate for this. The "high" scallop and potato rate shows a clear-cut reduction in the tendency toward foaming; the more food being fried, the more rapid the replacement of used fat with fresh fat.

Hydrolysis

This is the reaction of water from the food with the frying fat to form free fatty acids. The rate of hydrolysis or free fatty acid development depends on the following factors:

1. Amount of water released into the frying fat. The greater the amount of water, the more rapid the change. Water is generally introduced from the food that is fried. Fresh potatoes, for example, contain about 85% water. Potatoes that have been pre-blanched and frozen by large food processors contain over 50% water. Water can be introduced in other ways, such as during the kettle-cleaning operation.

2. The temperature of the frying oil. The higher the temperature, the more rapid the rate of free fatty acid production.

3. The rate of fat turnover is also important; with all other conditions being equal, the more rapid the replacement of used fat with fresh fat, the slower the rate of free fatty acid development.

4. The number of heating/cooling cycles of the oils.

5. The more crumbs and burnt food particles accumulate in the frying system, the greater the rate of development of free fatty

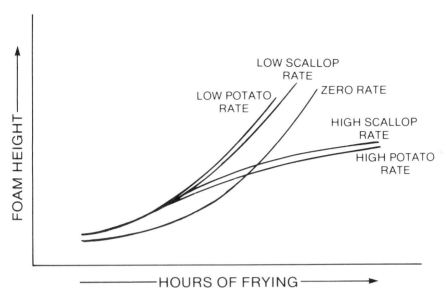

FIGURE 7.2.   Free fatty acid rise during deep fat frying.

acids. Therefore, proper and frequent filtration is important in keeping this effect to a minimum.

Fat quality plays a lesser role in the hydrolysis reaction than color formation, oxidation, and polymerization. Differences in free fatty acid levels in frying fats do not necessarily constitute a differentiation between good- and poor-quality fats in the frying system. Free fatty acid level does not correlate well with fried food quality. Furthermore, free fatty acids are somewhat volatile at frying temperatures.

The effect of increasing quantities of food and, thus, moisture is shown in Fig. 7.2. Heating alone—without frying food—does not produce much free fatty acid. Frying potatoes or scallops (both of which are high-moisture foods) at a "low" rate produces more free fatty acid than is produced at the zero food rate. Frying potatoes and scallops at a "high" rate produces significantly greater levels of free fatty acid. The amount of free fatty acid formed is directly proportional to the amount of steam released by the food into the fat. This, of course, occurs in both large food frying systems and in foodservice fryers.

Frying large quantities of high-moisture foods tends to increase the rate of free fatty acid development, but it must be kept in mind that frying greater amounts of any kind of food has a very positive effect on the life of the frying fat, because frying a greater amount of food increases the rate at which fresh fat is added. This keeps the frying fat in better condition, as measured by color, oxidative changes, and even free fatty acid changes. Therefore, care must be taken in trying to arrive at general conclusions without having the complete knowledge and facts associated with each frying operation that is being studied.

GOOD FRYING PRACTICES

Frying Temperatures

Most foods fry properly at a range of 325–375°F (163–191°C). Temperatures close to 400°F (204°C) ordinarily produce a browned surface before the inside is completely done. By the time the inside is properly cooked, the food is burned on the outside. Generally speaking, frying should never be done above 400°F (204°C). With modern, fast-temperature-recovery kettles, it is seldom necessary to fry above 350°F (177°C). In some food processor frying systems, a higher temperature may be justified where the time of food immersion is only 1–2 min. In addition, this type of system is usually aided further by continuous filter systems and very rapid frying oil turnover.

During slack periods, the temperature of the frying oil should be reduced to between 200 and 250°F (93 and 121°C) (or preferably turned off), since keeping the fat hot for long periods of time without cooking greatly accelerates oxidation. In food processing plants, where frying is generally continuous, the frying oil should be raised to frying temperature just before start of frying and should be turned off as soon as frying is completed. High temperatures cause the fat to oxidize, resulting in the earlier development of a tendency toward foaming, a definite darkening in color, and a noticeable increase in the fat's tendency to smoke.

With solid or plastic shortenings, special care must be taken when charging a kettle with fresh fat. Because solid fat absorbs and conducts heat slower than a liquid oil, a low heat should be applied to cold plastic fat when bringing it to frying temperature. Slow heating will avoid burning or scorching the fat nearest the source of heat.

A small amount of scorched or burned fat will accelerate the breakdown of all the fat in the kettle.

When using fluid shortening or liquid oil, the kettle may be filled to slightly below the "fill mark" and the thermostat can be set directly at frying temperature (350°F; 177°C). When the thermostat is turned on, the fluid shortening will not scorch, because it will flow evenly around the heating elements and completely cover them.

In view of the importance of proper frying temperatures, thermometers and thermostats should be checked regularly for accuracy—thermostats at least once per week and thermometers every 3 months.

It should be recognized that in most commercial food processors frying is done continuously, and the rate of breakdown is less rapid in such operations than in most foodservice operations that are intermittent. Examples of continuous frying operations include potato chips, corn chips, and precooked and frozen poultry and seafood.

Fat Turnover

Turnover refers to the amount of fat in the frying kettle that is replaced by fresh fat in a given time period (4). In commercial frying operations, such as for potato chips, doughnuts, and prefried frozen seafoods, it is not unusual to obtain 100% turnover of fat each day. The high rate of turnover is due to one or both of the following factors: (1) the foods being fried absorb large quantities of frying oil (e.g., about 40% of the finished weight of many potato chips is frying oil) or (2) there is a steady, continuous throughput of food during the entire time that the fat is at frying temperature. Based on nutritional considerations, a lot of work has been done to try to decrease potato chip oil absorption by better control over potato quality, surface area of potatoes, and frying conditions. In addition, fabricated potato chips contain less absorbed oil, generally 25–30%.

In foodservice kitchens the rate of turnover varies considerably, but the average is in the 30–45%/day range. The low rate of turnover is due to one or both of the following factors: (1) the foods being fried absorb lesser amounts of frying fat (generally in the range of 5–15% of the weight of the finished food) or (2) the frying is generally not continuous but has "peaks" and "lulls," which results in fat being under heat for long periods without any of the fat being removed by absorption.

Under most restaurant conditions, it is difficult to achieve sufficient absorption to avoid fat discard. However, most fast-food operations require discarding oil every 7–10 days. A very few operators with a very efficient frying operation who follow good frying procedures and have a very high volume of fried food are able to avoid throwaway. The basic decision on oil/fat throwaway should always be made on the basis of fried food quality. It is better to discard one day too early than one day too late.

Fat turnover can be increased or brought up to a maximum level for a given operation by using a kettle of the smallest capacity in keeping with adequate production requirements. By using the smallest possible amount of fat in relation to the amount of food being fried, turnover is kept at a maximum. As a rule, operations can generally be made more efficient if two 15-lb-capacity kettles are used instead of one 30-lb-capacity kettle—or two 30-lb kettles instead of one 60-lb kettle. This gives the foodservice manager greater flexibility in managing the deep-frying operation. On the other hand, the manager should consider fried food quality as being most important. If it is found that it is more difficult to maintain fry temperature in a smaller kettle it could result in some foods being too greasy.

THE FRYING FAT

It is important to use the proper frying oil or shortening for deep frying. Good fried foods cannot be prepared without using a frying oil of good quality. The frying oil is not only a medium for transferring heat from the kettle to the food, it is a food, itself. In deep-frying, it is necessary to have the proper amount of fat absorption in order to obtain the character and eating qualities desired, but excessive absorption results in decreasing fried food quality. A good frying oil should have the following features:

1. Must not impart any off flavors to fried foods; Must have a bland, neutral flavor
2. Long frying life, resulting in an economical frying operation
3. Ability to produce an appetizing, golden brown, nongreasy surface on the fried food during its frylife
4. Minimize smoking after continued use
5. Ability to produce foods with excellent taste and texture

6. Minimize gumming, which aids in keeping equipment clean
7. Minimize oxidative changes
8. Uniformity in quality
9. Ease of use, including, both the form (fluid shortening is easier to use than solid shortening) and the package

Methyl silicone is generally used to minimize foaming. If the proper methyl silicone is used at the proper levels (usually 2–6 ppm of dimethyl polysiloxane) and added near the end of the processing cycle, it will help minimize foaming and oxidative breakdown at frying temperatures.

Fluid shortening can be poured or pumped, which saves considerable kettle or fryer filling time as compared with solid shortening. To load a kettle or frying system properly:

1. Underfill the kettle (because as fluid shortening heats it expands), but make sure coil or heating tubes are covered.
2. When the frying temperature is reached, add just enough shortening to bring the kettle to proper frying level. Do not make the mistake of adding a quarter of a can or so just to finish off the can. Overfilling could result in excessive splashing of very hot oil on frying personnel and on floors adjacent to the frying area.
3. Periodically add shortening to maintain the frying level as frying takes place.

To keep the frying fat and the kettle clean:

1. Filter at least once a day. Many foodservice operations use inline filters, and almost all commercial operations have a continuous or semicontinuous filtering system.
2. Wipe out kettle after filtering and before pouring shortening back.
3. Add more shortening to bring the frying fat up to the right frying level.

Figures 7.3–7.6 show examples of typical foodservice frying equipment. Good equipment is available in a large range of sizes in both floor and counter models and may be either gas or electric fired.

FOOD PREPARATION

Effect of Food to Be Fried

The type and condition of the food to be cooked affects the frying life of the shortening.

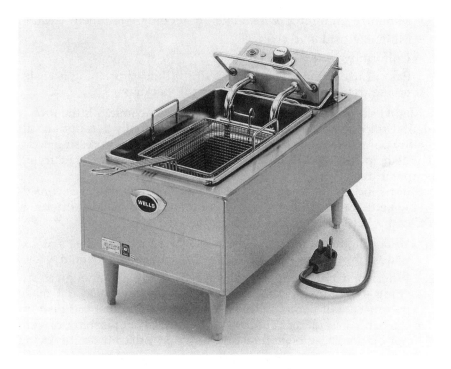

FIGURE 7.3.   Small electric counter fryer. (Courtesy of Wells Manufacturing Company.)

Seafoods contribute to rapid color darkening and oxidation of frying fats because of minor or trace elements present in seafoods as well as the breading and battering materials used. French doughnuts, which contain ammonium carbonate or ammonium bicarbonate as a leavening agent and which are high in egg content, cause excessive darkening of frying oil or fat.

Improper preparation of food frequently hastens frying fat changes. An excessive amount of moisture on the surface of food will produce violent bubbling and more rapid development of free fatty acids; therefore, excess moisture should be removed before the food is placed in the deep fat.

To ensure a smooth surface for frying, excess bread or cracker crumbs should be removed before a food is placed in the frying fat. Foods should never be added to the frying basket while the basket is over the frying fat. This minimizes contamination of the fat by

FIGURE 7.4. Electric floor-mounted fryer. (Courtesy of the Hobart Corporation.)

FIGURE 7.5.   Gas-fired floor model kettle with computer. (Courtesy of the Frymaster Corporation.)

loose crumbs. Loose crumbs or breading materials remain in the fat and become charred from prolonged heating. If the charred particles adhere to the fried foods, they can contribute poor flavors and black spots (the particles may look like pepper). They also accelerate color darkening, oxidation, and hydrolysis of the fat.

Food particles can be removed from the fat by straining through filter cones (Fig. 7.7). The fat should be kept warm at a temperature

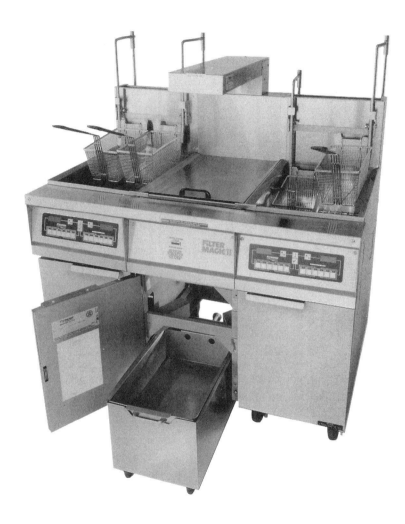

FIGURE 7.6.　Installation of multisized floor model fryers with filter system. (Courtesy of the Frymaster Corporation.)

FIGURE 7.7.   Straining through filter cones.

of about 150°F (66°C) or above. If the temperature of the fat is too low, filtering will be sluggish; however, at too high a temperature, there is greater danger that the operator will be burned by splashing fat and the fat can be oxidized, with a resultant reduction in fry life.

To ensure uniformity of finished foods, food pieces of similar size should be fried at the same time. In addition, deep-fried foods should be served immediately after frying, when they are at their flavor peak.

TYPES OF DEEP-FRYING EQUIPMENT

The proper use and care of the proper frying and filtering equipment are very important factors in obtaining the most efficient operation and producing top-quality fried foods. Fryers should be of the correct size and installed in the proper numbers. Other factors to consider are (1) the source of fuel, (2) the speed of temperature recovery, (3) the ability to conserve fuel, and (4) safety.

Foodservice fryers are either counter models mounted on top of work tables or floor models mounted directly on the floor (Figs. 7.3–

7.6). Counter fryers are generally sized to hold 10–30 lb of frying shortening. Some later models have added features, such as automatic timing and/or pop-up baskets for the frying of various foods. Many fast food operations utilize push-button computerized cooking controls for the various types of foods that are deep fried. Inline filters are becoming more prevalent in helping keep the frying fat in better condition.

Kettles that are mounted on the floor vary in frying shortening capacity from 30–60 lbs. Some models have a capacity of up to 120 lb, but these larger kettles are generally for specific specialty frying purposes (e.g., doughnut frying).

With any electric fryer it is extremely important to match the voltage supplied to the restaurant unit with the voltage rating of the fryer (e.g., 208, 220, or 240 V; 1 phase or 3 phase).

PROPER USE OF FRYING EQUIPMENT

Modern deep-frying equipment made by reputable manufacturers is easy to use and generally easy to maintain in good condition. In this section, suggestions are made on how to get the best results with frying equipment.

Avoid metal contamination, especially with copper. The presence of very small quantities of copper in a frying fat can shorten frylife by 20–30%. It is, therefore, necessary to avoid the contact of hot frying fat with copper or copper-bearing alloys, such as brass. Most manufacturers are aware of this and have stopped using such metals in the kettles or baskets. Occasionally, copper valves are encountered because they are less expensive than stainless-steel valves. Sometimes repairs are made with copper-bearing materials. When frying baskets are repaired, copper welds may be used. Obviously, this is a bad practice. Some thermocouples (the sensing element of the thermostat) are chrome-plated copper, and as the chrome plate wears off, copper is exposed. All such copper should be replaced to improve frylife.

Avoid overheating. Both gas and electric kettles are equipped with thermostats to regulate the temperature of the frying fat. When the temperature sensed by the thermocouple is below the setting on the thermostat, the heat is on. When the sensed temperature is above or equal to the set temperature, the heat is off.

It is important to remember that the heat is either off or on, and regardless of the size of the differential between the set and the

sensed temperature, the kettle will not heat faster. Many fry operators think that if they set the thermostat at 400°F (204°C) the kettle
will heat up faster than if set at 350°F (177°C). The usual result is
that they forget to turn the temperature back and cause a more rapid
breakdown of the fat.

In addition, the accuracy of the thermostat should be checked at
least once a week with an accurate thermometer.

Another possible problem is the operator who wants to fry more
food in less time and overloads the kettle with food, sometimes raising the thermostat setting to 375°F (191°C) or 400°F (204°C) in an
attempt to compensate. Once the thermostat comes on, the kettle
is delivering all the heat energy it is capable of delivering, so if the
temperature is below 350°F (177°C), the system will not deliver more
heat whether the thermostat is set at 350°F (177°C) or 400°F (204°C).
If the energy requirements of the food are more than the energy
output of the kettle, then the temperature will continue to drop.
Thus, overloading always leads to longer cooking times, more fat
absorption, and less appetizing foods. The food most commonly associated with such problems is potatoes because of the high moisture content and the tendency to pack into the baskets. Even the
newest fryers can be overloaded with potatoes. So it is important
to standardize loading practice to prevent this. The frying fat should
be back to its original temperature in less than 1–2 min after the
previous load of food is removed from the fryer. If not, the size of
the load should be adjusted downward until the fryer does recover
that quickly. In the very large commercial frying systems that continuously convey the food through the hot oil, temperature drops
are seldom a problem.

Electric kettles can become a fire hazard if the level of fat in the
fryer falls below the top edge of the heating coils. The temperature
of the exposed coils can reach the spontaneous ignition point of the
frying fat and cause a fire. In addition, this exposure contributes to
considerably greater oxidation and polymerization of the frying fat
and can result in premature frying fat breakdown. The scorched fat
can also result in the fried food having a scorched flavor.

Gas kettles are often operated with their pilot lights on 24 h a
day. If the pilot light is low, this is not a problem. However, in cases
where the pilot light is set high or the design is such that the pilot
light burns against the kettle, then temperatures of about 300°F (149°C)
may be maintained in the kettle even during "off" periods. This can
lead to more rapid oxidative breakdown of the fat.

All fry kettles that fry breaded foods should be cleaned daily. This is true regardless of the size of the kettle and whether it is the food processor or foodservice equipment. In some commercial food processor operations such as potato chips or other nonbreaded or nonbattered foods, a once-a-week cleaning may suffice.

It may be necessary to scrub kettles and baskets occasionally with a washing compound and boil them out with a good low-sudsing detergent. Usually this is done once a week or at the time of discard of the frying oil. About 2 oz. of detergent should be used for each gallon of water used for boiling—about 4 oz. of detergent for a counter frying kettle with a 15-lb fat capacity. One half to one ounce of detergent per gallon of water should suffice for food processor kettles.

A low-sudsing detergent is recommended for kettle cleaning. Harsh alkalis can cause damage to the kettles and increase the hazard of harmful residues remaining. Kettles should be boiled for 15–20 min to remove the gums and polymers that form with any frying fat in normal use. Brushing should accompany this type of cleaning. Soft brushes are preferred to stiff abrasive brushes, which can rub the coating off heating elements. After cleaning, the kettles and baskets should be thoroughly rinsed three or four times to make sure that all traces of detergents are gone. If they are not rinsed thoroughly, hydrolysis may be accelerated and, in extreme cases, the flavor of the fried food may be impaired.

Many fry operators use a vinegar rinse to help neutralize an alkaline detergent. The value of such a practice is debatable because it might be used to cover up poor rinsing practices. Besides, it is still desirable to rinse out the vinegar residue before the frying fat is returned to the kettle. After thorough rinsing, the kettle should be dried before the fat goes back into the fryer. If excess water is left in the kettle when the fat is brought up to frying temperature, very violent bubbling may occur, which can be hazardous, and the rate of hydrolysis will increase at an excessive rate.

In commercial large-quantity frying, the basic principles outlined in the preceding pages apply. The basic differences are considerably larger frying systems and the continuous frying of food. By using the fundamental principles of good frying practice, along with common sense, the result will be an efficient commercial operation.

Troubleshooting

The following problems may occur when using deep fat frying methods. Symptoms, causes, and solutions are given.

Frying Oil Darkens Excessively

Symptoms:

Oil or fat darken prematurely

Possible Causes

1. The use of an inferior frying oil or adding rendered drippings, poultry fat, and so on to kettle or frying system
2. Overheating of fat, possibly caused by a faulty thermostat
3. Inadequate filtering of oil or fat
4. Inadequate or improper cleaning of equipment
5. Hot spots in kettle or frying system
6. Frying improperly prepared food

Probable Cures

1. Use a good-quality stable oil or fat. Never add rendered drippings. Avoid frying of sausage and/or bacon. In the larger food processing plants, watch for the excess rendering out of fat from foods being fried, for example, tallow from meat patties.
2. Fry most foods at 350°F (177°C). Make sure thermostat setting is correct.
3. Filter oil at least daily and more often if a lot of breaded foods are fried.
4. Clean equipment (boil out and scrub) at least once a week, making sure to wash out the smallest trace of detergent.
5. In gas-fired kettles, make sure flame in firing tubes is evenly adjusted, with no thin spots on tube, and that all heated surfaces of the kettle are covered with oil. If using a solid fat, make sure it is melted before turning heat on full.
6. Make sure that hinged immersible electric heaters are fully immersed when heating fat, as a sprung hinge can sometimes cause the front portion of the heating element to be raised slightly.
7. Make sure foods to be fried are not of such a nature that they cause excessive darkening—for example, potatoes subjected to too strong a treating solution (sodium bisulfite) will darken the fat at twice the normal rate.

8. Make sure no foreign matter is in the frying oil, such as detergent left in the kettle after washing or crumbs accumulated during frying. Foods should not be salted directly over frying fat. Trace elements in the salt can cause early darkening.

Frying Fat Smokes Excessively

Symptoms:

Noticeable smoke coming from the kettle or frying system, sometimes filling the room with a slightly blue haze.

Possible Causes

1. A high level of free fatty acid in the frying oil or fat
2. Inadequate filtering of oil
3. Use of improperly prepared food, especially breaded or battered items
4. Use of inferior oil or fat
5. Overheating, possibly caused by faulty thermostat
6. Inadequate or improper cleaning of equipment
7. Hot spots in frying system
8. Poor ventilation

Probable Cures

1. Filter fat at least daily—more often if needed. Employ continuous filtration in large food processing plants that are frying breaded/battered foods.
2. Make sure foods to be fried are not too heavily or too loosely breaded and that excess loose breading is shaken from food before the food is placed in the frying fat. Excess smoking may be coming from the burning crumbs instead of from the oil.
3. Use the best-quality most-stable oil. Never add rendered drippings. Do not deep fry bacon or sausage. In large plants, watch for excessive rendering out of tallow and poultry fat from foods being fried.

4. Fry at a proper temperature, usually 350°F (177°C). Make sure thermostat is correct.

5. Clean equipment (boil out and scrub) at least once a week, being sure to wash out every last trace of detergent.

6. Make sure flame in tube is evenly adjusted, with no thin spots on tube, and that all heated surfaces of the frying system are covered with oil. If using a solid fat, make sure it is melted— to carry heat away by circulation—before turning heat on full. Make sure that hinged immersible heaters are fully immersed when heating fat, as a sprung hinge can sometimes cause the front portion of the heating element to be raised slightly.

7. Make sure ventilation is adequate. Sometimes this makes the difference between "excessive" and "normal" smoking.

Life Gone from Fat

Symptoms:

Frying oil will not hold heat or brown the food; foods do not cook and/or brown properly in expected time.

Possible Causes

1. Excessive foam development
2. Overloading kettle
3. Too low a frying temperature, possibly caused by a faulty thermostat
4. Improper preparation of food
5. Lack of proper kettle recovery
6. Not cooking long enough

Probable Cures

1. Make sure the frying oil is not foaming. Foam is the minute bubbles that creep up the sides of the frying machine.
2. Make sure the initial drop in frying temperature due to frying load is not excessive.

3. Make sure food is properly drained of excess moisture. Be sure that the food is properly prepared; choose breaders or batters that brown properly within the proper cooking time.
4. Make sure kettle recovery time is sufficient to cook the load placed in it. Do not overload!
5. Be sure cooking time is correct. Use a timer.

The ability of a fat to brown food depends on the frying time, temperature of the fat, and the color of the fat.

Frying Fat Foams Excessively and Prematurely

Symptoms:

Fine white or yellowish white bubbles rise up the sides of the kettle, and these bubbles remain for a prolonged period.

Possible Causes

1. Use of inferior oil or fat
2. Overheating of fat, possibly caused by a faulty thermostat
3. Hot spots in frying equipment
4. Holding fat idle at frying temperature for long periods of time without frying food, resulting in poor frying fat turnover
5. Kettle too large for the amount of deep-frying
6. Improper care of frying oil or fat

Probable Cures

1. Use a high-quality oil or fat. Never add rendered drippings.
2. Fry at proper frying temperature, usually 350°F (177°C); do not fry above 375°F (191°C), especially for small operations. Make sure thermostat setting is correct, regardless of the size of the frying operation.
3. Make sure flame in firing tubes is evenly adjusted, with no thin spots on tube, and that all heated surfaces of the kettle are covered with fat. If using a solid fat, make sure it is completely melted before turning heat on full. Make sure that hinged im-

mersible heaters are fully immersed when heating fat, as a sprung hinge can sometimes cause the front portion of the heating element to be raised slightly.

4. Turn frying kettle down to 250°F (121°C)—or completely off—when not in use for 15 min or longer. During off-peak hours, you may be able to operate with only one frying kettle. This will save fat. In large food processor frying it is especially important to not keep a very large frying machine at elevated temperatures when food is not being conveyed through the system. This wastes energy and breaks down the oil or fat.

5. In many cases, it is best to have two small kettles rather than one large kettle. This will permit turning one kettle off when not needed.

6. Filter frying oil at least once daily—more often if needed.

Fried Foods Are Greasy

Symptoms:

Fat drips from finished fried foods. Fried foods are grease soaked.

Possible Causes

1. Frying at too low a temperature.
2. Overloading the kettle with food, or overloading conveyor belts with food.
3. The frying fat is foaming.
4. Improper preparation of food.
5. Keeping food in frying oil after it is cooked.
6. Improper draining of food after frying.
7. Slow temperature recovery.

Probable Cures

1. Make sure the fat is up to temperature at the start of frying so that it is still above 320°F (160°C) after initial loading.
2. Make sure initial drop in frying temperature due to frying load is not excessive. Do not overload.

3. Make sure food is frying in fat, not foam. If oil or fat is foaming, it must be discarded.
4. Excessive water will result in an excessive temperature drop and may result in greasy fried foods.
5. When food is done, remove immediately from the frying oil or fat and allow to drain.
6. After frying, allow excessive fat to drip into the kettle for 15–30 s.
7. Make sure the frying fat recovers to the normal starting temperature before frying additional foods.

The absorption rates of all frying fats and oils are about the same. The fats that melt at higher temperatures, such as the hydrogenated shortenings, sometimes appear to be absorbed less in the finished food (as compared with fluid frying shortenings) because, as the finished product cools, these higher-melting fats congeal and give the food the appearance of being drier or less greasy. However, analytical tests generally show no difference in actual fat content.

Fat Has "Objectionable" Odor or Flavor

Symptoms:

Odor of the frying oil or fat is "unappetizing."

Possible Causes

1. Use of an inferior fat, or the fat has deteriorated because of abuse. This includes the addition of rendered drippings
2. Use of poor-quality food
3. Excessive crumbs in the fryer
4. Fat dripback from the stack or overhead hood, or other foreign material getting into kettle

Probable Cures

1. Use a high-quality frying oil. Do not add rendered drippings. Discard when it develops a disagreeable odor.

2. Check the odor of the foods to be fried.
3. Strain regularly to remove burnt materials.
4. Make sure stacks are properly cleaned, and keep frying systems covered when not in use.

Fat Gives "Objectionable" Flavor to Fried Foods

Symptoms:

Fried foods have "objectionable" fatty flavor.

Possible Causes

1. Use of an inferior fat, or the fat has deteriorated through abuse. This includes the use of rendered drippings or the frying of bacon or sausage. In large commercial operations, be extremely alert as to the amount of fat rendered out of foods such as meat patties and poultry.
2. Dripback from exhaust stack into fat.
3. Use of poor-quality food.
4. Inadequate filtration of crumbs from the frying oil or fat, causing burned crumbs to build up.
5. Excessive fat absorption.
6. Poor turnover.

Probable Cures

1. Use a high-quality oil or fat. Do not add rendered drippings. Discard when the fat develops a disagreeable odor or foams excessively.
2. Make sure exhaust stack system is so constructed as to minimize dripback from the stack into the kettle. Clean stacks and stack filters as needed—usually once a week.
3. Check the flavor of the foods to be fried.
4. Filter frequently; do not allow burnt crumbs to build up and impart off flavors to food.

5. Make sure the frying temperature is correct and that baskets are not overloaded.

6. Add fresh oil or fat frequently, keeping it up to the manufacturer's suggested frying level in the kettle. Good turnover means better-flavored fried foods and a longer frying life.

7. Keep kettles covered when not in use.

FILTRATION OF OILS AND FATS

Definition and Purpose of Filtration

The purpose of filtration is to remove crumbs and sediment (some of which is very fine) from the kettle before these solids can harm the shortening or the appearance and flavor of fried foods. Some of the problems that can be caused if these solids are not removed include excessive and/or premature darkening of the shortening, off-flavor development, burnt flavors, poor color and appearance of the fried food, and excessive smoking of the shortening at frying temperatures. Proper filtration helps to ensure good-looking fried foods, prolongs the life of the shortening, and cuts the overall operating costs of the frying operation.

Methods of Filtration

Gravity Filtration (for Relatively Small Operations)

This is simply filtration via a filter paper or cloth, used with or without a filter aid. (See Fig. 7.7.) For the most efficient filtering, the frying oil or shortening must be at least 150°F (66°C) to ensure complete liquefaction and to lower the viscosity of this oil so that it can pass freely through the filter. If a filter aid is used, it is usually stirred into the hot fat before the shortening is poured into the filter paper. As the shortening passes through the filter paper or cloth, the filter aid forms a layer on the surface of the filter medium. As the layer grows thicker, the filtration becomes more effective until finally the layer becomes too thick and slows down the rate of filtration.

Examples of gravity filtration are a rack and cone system and filter bags. The latter are simply porous bags that are attached to the outlet pipes of a fryer so that the shortening can drain into a holding

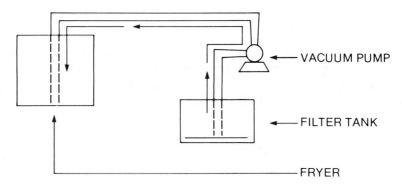

FIGURE 7.8.   A simplified vacuum filtration system (illustrating a one-way pumping system).

container. Gravity filtration is usually trouble-free, as there is no machinery to maintain. However, this type of filtration is slow and usually not practical for use with fryers with capacities over 30–35 lbs.

Mechanical Filtration

There are two types of mechanical filtration: vacuum and pressure. Both of these systems use a pump to speed up filtration. Vacuum filtration uses a vacuum pump to pull the shortening through the filter. A pressure system uses a positive displacement pump to push the shortening through the filter. Block diagrams illustrating the operation of both the vacuum pump and the positive pressure pump methods of filtration are shown in Figs. 7.8 and 7.9. Most foodservice operations that employ mechanical filtration use the vacuum pump type. The positive pressure system is generally used in commercial food processing operations.

There are two basic types of vacuum filter. One type consists of a receiving vessel (or pot), the filter, and a one-way pumping system that will return the filtered shortening to the fryer. The second type of vacuum filter has a pumping system that is capable of withdrawing the shortening from the fryer as well as returning the filtered shortening.

One advantage of being able to both withdraw the hot shortening from the kettle as well as return the filtered shortening to the kettle by means of the filter is the safety in handling hot shortening. This

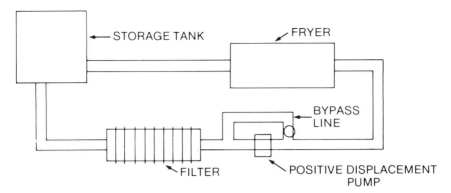

FIGURE 7.9.   A simplified mechanical pressure filtration system.

becomes more important when the shortening capacity is greater than 30 lbs.

The operation of these vacuum filters is very easy and usually consists of the following steps:

1.   Drain or pump the shortening into the receiving tank.

2.   Loosen the crumbs and residue from the kettle walls using a spatula or knife.

3.   Wipe out the kettle, or, if the kettle is a large floor model, a small amount of shortening may be pumped from the filter and the stream of hot fat can be used to wash down the fryer walls. This shortening plus crumbs should be returned to the filter tank. Great care must be taken not to aerate the shortening while washing down the fryer walls.

4.   Wipe off the coils or heating tubes.

Positive-Pressure Filtration

This system uses a positive displacement pump. The hot shortening is drained from the fryer to the pump, pushed by the pump through the filter, filtered, and returned to the fryer or to a storage vessel. This type of filtration lends itself best to large-scale frying operations.

The bypass line around the positive displacement pump is necessary because if the filter were to clog up during filtration, a very dangerous back pressure would build up. This bypass line is a very necessary safety device; however, it can cause problems because if the bypass line does not contain a meter to indicate flow, the operator cannot tell whether his shortening is going through the filter or merely being pumped around it. This situation could result in a large portion of the shortening not being filtered.

Figure 7.10 illustrates a good continuous centrifugal filtration system.

Summary of Steps Necessary for Good Filtration

1. Warm the shortening until it is at least 150°F (66°C).
2. Turn the kettle off before starting to filter. The heating coils should be off when not covered with shortening.
3. Empty the frying fat into the filter.
4. Scrape down the fryer walls with a spatula to loosen crumbs and residue. A stainless-steel pad may be used, but never use copper or brass scouring pads.
5. Use a small amount of filtered shortening to rinse down the fryer walls and drain the shortening back into the filter.
6. Fill the fryer with filtered shortening.
7. If a mechanical filter is used, never splash the hot shortening carelessly around in the fryer—fat aeration and oxidation will be accelerated.
8. Turn the fryer on only after the shortening has been returned to the fryer and the coils are covered.
9. Add fresh shortening to the kettle to bring it up to the proper level.

Filter Aids

Filter aids are added to a filtration system either to improve clarity of the shortening or to increase the rate of filtration.

In order to do these jobs, a filter aid must form a porous cake, have a low shortening-absorption level, and have good particle size distribution. Too-fine particles yield slow filtration, whereas too-coarse particles yield too-fast filtration.

FIGURE 7.10.　Continuous centrifugal filtration system. (Courtesy of Heat and Control, Inc.)

There are hundreds of different brands of good filter aids on the market today, many of which are packaged by the manufacturer of the filters. Most filter aids are some type of diatomaceous earth.

Filtration Troubleshooting

Filter Problems and Causes

PROBLEM	PROBABLE CAUSE OR CAUSES
1. Black specks on fried foods.	Not filtering, a leak in the filter medium, or not filtering often enough.
2. Breading burning between heating coils.	Not filtering often enough to keep breading and crumbs away from coils.
3. Localized large bubbling of frying oil while kettle is just heating	Excess water in the frying system that is bubbled off when heat is applied.
4. Frying oil tastes or smells "scorched" or "burnt."	Breading and crumbs have burnt in the frying oil. There is no alternative but to discard the oil if this is allowed to occur.
5. Mechanical filter will not pump the frying oil.	The filter pump line may be clogged, or the filter pump motor may not be working, or the frying oil may be too cold.
6. Frying oil comes from filter slowly and is mixed with air.	The filter paper may be clogged, the oil may be too cool, or the fitting on the return hose may be loose, allowing air to be drawn into the hot oil.
7. Crumbs and/or filter aid are found in filtered oil.	Filter medium has a hole in it. Replace paper or cloth and filter aid.
8. Filtered oil is not clear.	The particles to be filtered out may be very small and the filter medium may not be capable of removing particles of this size. The use of filter

aid plus the correct filter medium usually solves this problem. A finer grade of filter medium (smaller pore size) may be a simple answer.

9. Filtered oil has "earthy" taste or smell.

Change brand of filter aid. Some very poor-quality brands have been found to impart off flavors to the oil.

10. Coils of the fryer smoke excessively during filtration.

Heating coils should be turned off during filtration.

PROCEDURES FOR FRYING MAJOR FOOD ITEMS

French Fried Potatoes

Almost all french fries used in foodservice have been preprocessed and frozen in large commercial operations. As potatoes are the most important single food that is deep-fried, a knowledge of the science of potatoes should be of value to many food technologists and food-service managers.

The typical composition of fresh uncooked potatoes is as follows: about 85% moisture, 0–1% fat, 13–18% starch, 1–4% sugar, and 1–2% protein. Good-quality fresh potatoes are a prime requirement for making good-quality french fries. The potato must be fried to a golden brown color with a moderate crust, a soft mealy center, and a good potato flavor.

Selection, Preparation, and Storage of Potatoes

Selection of Fresh Potatoes

The general rule is to select potatoes with high solids (high specific gravity/mealy varieties) rather than low specific gravity (the non-mealy varieties). The mealy varieties yield the best eating qualities and the most attractive color and appearance in the finished french fries. The varieties most commonly used include Irish Cobbler, Russet Burbank, Russet Rural, Sebago, and Kennebec.

Storage of Potatoes

Fresh potatoes should be stored at 50–60°F (10.0–15.6°C). Storage temperatures below 50°F (10.0°C) will cause a chemical reaction of the starch in the potato—some of the starch changes to sugar, and when the potato is fried it darkens excessively due to the carmelization of these sugars. This chemical conversion takes place slowly, showing its greatest effect in 7–10 weeks. Most potatoes stored 8 weeks at 40°F (4.4°C) are unsuitable for producing golden-colored french fries. However, it is possible to reverse this reaction by storing the potatoes at 60–70°F (15.6–21.1°C) for several weeks, as these temperatures favor the conversion of some of the sugars to starches. However, at temperatures above 60°F (15.6°C), other deleterious effects are promoted, such as sprouting, molding, and rotting. Therefore, both storage time and storage temperature can be critical in producing the best-quality french fries.

Preparation of French Fries

Proper peeling is important to remove all traces of skin, the eyes, and blemishes from the potato before slicing into strips. Slicing of strips should be done in a manner that will ensure uniform slices. When pieces of widely varying thickness are fried simultaneously, there will be a big difference in the quality of the french fries produced.

Soaking the sliced potatoes in cold water will prevent the potato pieces from turning brown and will help to keep the pieces crisp. After soaking, care should be taken to drain or shake the potatoes to remove as much water as possible before frying.

For the few foodservice operators and foodservice commissaries that prepare their own french fries, a discussion of blanching/ browning will be helpful. It also serves to better understand all phases of potato preparation.

Blanching is a technique used with fresh potatoes to permit faster preparation to the finished fries at the time they are needed in a hurry. The preblanched potatoes may be held for several hours at room temperature and fried to doneness in about 2 min. A common belief has been that blanching should be carried out at a temperature well below the browning temperature. This is a fallacy. Blanching and browning at the same temperature yields good-quality french fries with lower fat absorption. Potatoes blanched at 300°F (149°C) and fried at 350°F (177°C) usually contain 3–5% more absorbed fat than potatoes both blanched and fried at 350°F (177°C). The optimum blanching conditions seem to be blanching at 350°F (177°C) for 2–3 min and browning for an additional 2–3 min at the same temperature. This varies, of course, depending on the size and type of potato.

Frozen Preblanched French Fries

This has become the most important type of french fries in use today. Storage of commercially prepared frozen french fries consists merely of keeping the product frozen. These prefrozen french fries yield the best finished product when fried without defrosting. After defrosting, these potatoes should never be refrozen.

Almost all commercially prepared frozen french fries are pre-blanched—some in tallow, some in palm oil, and some in partially hydrogenated soybean oil. This preblanching cuts down on the frying time necessary to obtain a thoroughly cooked, golden french fry in the foodservice operation. The preblanched potato usually contains 5.0–7.0% fat, whereas the raw white potato contains 0–1%. When the preblanched potatoes are fried, they usually contain a total of 12–16% fat. Again, this depends to a considerable extent on the size and shape of the potato.

Some potatoes processed in this manner are coated with a dextrose solution in order to obtain more even browning of the french fries. When this process was first developed, problems evolved with the use of too strong a dextrose solution and reduced frylife of the frying oil. However, this method appears to work very well now.

Deep-frying Procedures

1. All frying should be done at 350°F (177°C). This includes blanching as well as browning.
2. Shake off all excessive moisture before starting the frying.
3. A fat-to-product ratio of about 10 lbs of fat to 1 lb of potato should be used.

4. Remember to skim the kettles often. In general, the thinner the cuts, the more necessary skimming becomes.

5. Do not let the potatoes sit in the baskets over the hot shortening after frying. Prolonged exposure to the heat produces a limp potato.

Keeping the Fried Product

In fast-food operations, french fries must be cooked ahead of demand. A few simple practices will help the french fries keep their appearance and eating quality:

1. Remove the french fries from the fry baskets immediately after frying.

2. If warming lamps are not used, the product should be served within 5 min.

3. Using warming lamps, the french fries may be kept for 10–15 min. Because the warming lamps dry the french fries out and produce a crust, it is important to check the eating quality of the product often during the holding time.

4. Redipping the product (a 1–2-min fry to rewarm the product) greatly increases the fat content of the potatoes, which produces a poor-quality, greasy french fry.

Seafoods

Varieties of Seafood

There are two major types of seafood: finfish and shellfish. Both types of seafood can be purchased fresh or frozen. The commercial food processor provides a wide range of frozen fresh seafoods, battered or breaded uncooked products, and prefried/precooked seafoods.

In foodservice outlets, fresh fish should be stored below 40°F (4.4°C) until ready to use. The use of crushed ice is generally preferred to dry refrigeration. Frozen seafoods should be kept frozen until ready for use. This is critical to ensure freshness and the serving of high-quality prepared seafoods.

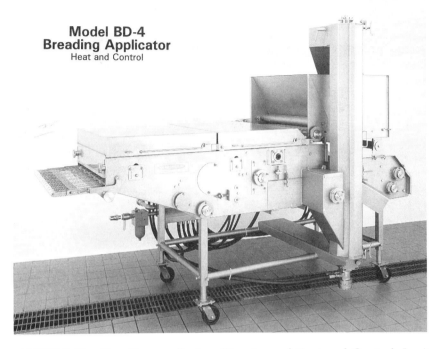

**Model BD-4
Breading Applicator**
Heat and Control

FIGURE 7.11.   Breading applicator. (Courtesy of Heat and Control, Inc.)

Breading of Seafoods

In medium to large seafood processing plants the precoating is done by automatic breading and/or battering machines (Figs. 7.11 and 7.12). In some foodservice outlets, the breading and/or battering is still done by hand. Regardless of the size of the operation, the basic principles are the same. In large operations, prepared breading and prepared batter mixes are available to meet every possible need. Fast browning breaders are often used with seafoods that are conveyed through the frying in a short time, for example 1–2 min. In addition, they may be flavored and colored to meet specific needs. With batter mixes, the frying operator generally just adds water or has equipment that meters the water into the dry batter mix. These batter mixes will also vary in flavor and color to meet specific needs. The use of buttermilk solids is often used in both breaders and batters for flavoring. Some batter mixes will contain leavening agents, if it is desirable to have the batter expand when it comes in contact

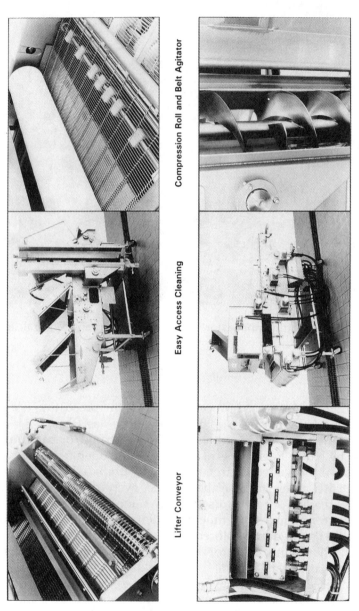

FIGURE 7.12.   Components of breading applicator. (Courtesy of Heat and Control, Inc.)

with the hot frying oil. In small operations, especially small food-service establishments, batters are made from simple mixtures of egg and water. They vary from 1 to 3 dozen eggs to a gallon of milk, buttermilk, or water, depending on the desired consistency.

Some seafoods, such as oysters or clams, may require two bread-ings to obtain the proper coating. If the seafood is double breaded, it should be dipped in a wash or batter between breadings.

Important Breading Precautions

The object of breading is to coat the food uniformly with a breaded coating that will form a crisp covering when the food is fried.

A uniform, dry coating is important in obtaining an evenly browned seafood that has minimum fat absorption.

Cracker crumbs or meal, corn flour or meal, and seasoned flour produce a slower-browning food. These coatings should be used on seafoods requiring a long frying time.

Bread crumbs are generally used on seafoods that require quick cooking of 1 or 2 min.

Foods should not be breaded too far in advance because the breading will become soaked and this will result in uneven brown-ing. When the breading becomes soaked, the food should be re-breaded; however, rebreaded seafoods tend to brown faster than freshly breaded items. .

Deep Fat Frying

Virtually all seafoods can and are deep-fried by foodprocessors and foodservice outlets. Very large pieces of fish are usually cut into small portions to ensure thorough cooking without excessive browning. If pieces of fish having great differences in thickness are to be deep fat fried together, the larger pieces must be started before the smaller pieces. In food processor operations, all seafoods, in-cluding fish sticks, are generally well controlled as to size.

Frying Procedures

1. Remove the breaded seafood from the refrigerator, shake lightly to remove excess crumbs and breading, and place in the fry bas-ket. In food processor operations, the breader and/or batter ap-

plicating equipment must be preset, and the conveyors should have methods of removing the excess before the food is conveyed through the frying oil.

2. Deep fat fry at 350°F (177°C) until golden brown.

3. Gently shake the fry baskets early during frying to make sure that the seafood pieces do not stick together. Large frying systems generally provide sufficient agitation to accomplish this purpose.

4. After frying, allow the seafood to drain in the baskets over the fryer for 5–10 seconds to allow excess shortening to drain off. In large plants, conveyor belts are again designed for drainage before freezing or packaging.

5. Remove the fried food to a plate and serve immediately.

6. Skim and filter the frying oil regularly—burnt breading will accelerate oil breakdown and adversely affect the flavor of fried foods.

7. All frozen seafoods must be fried directly from the freezer. If frozen foods are permitted to thaw, the fat absorption will greatly increase and the food will not be crisp.

Seafood Sanitation

Proper sanitation procedures for all food plants and foodservice outlets are coming under vastly increased scrutiny by regulatory officials. Even when not under regulatory control, good sanitation procedures must assume a high priority wherever foods for human consumption are prepared or stored. Only one outbreak of illness that can be traced to any plant will result in an immense loss of business, or even put a plant or foodservice outlet out of business. Foods that utilize batters or breaders containing eggs are especially vulnerable.

The following are the major areas of concern in the sanitation efforts of a food processing plant or a restaurant outlet.

Receiving: Frozen foods should arrive frozen and be put into the freezer immediately. Fresh seafoods should be received and held iced until used.

Refrigeration area: Foods should be stored at 32–40°F (0–4.4°C) for refrigerated items and at temperatures below 0°F (−17.8°C) for frozen foods. They should be covered and rotated. The floor of the cooler area should be made of tile so that it can be thoroughly cleaned and hosed down. Store all seafoods at least 6 in. off the floor.

Breading area: Equipment in this area must be cleaned carefully to make sure that bacteria are eliminated. The egg/milk wash used in this area provides an excellent growth medium. A strong sanitizing solution should be used regularly.

Refrigerator: Breaded items should be held at refrigerator temperatures until used. Frozen items that have been defrosted for use should be kept in the refrigerator.

The fryer and broiler areas generally do not contribute bacteriological problems due to the high heat used; however, tops, sides, and edges of equipment are good breeding grounds for bacteria. The floor must be cleaned regularly to remove food material.

Serving equipment: Any piece of equipment that touches the food must be kept clean. Unnecessary contamination of food often occurs from unclean serving tools or dirty hands.

Employee dress and habits: Cleanliness is an absolute necessity. In this and other sanitation/cleanliness procedures, always follow Good Manufacturing Procedures.

Oil/Fat Content of Some Seafoods

FOOD	%OIL/FAT
Shrimp, raw, breaded	1
Shrimp, fried	12
Clams, raw	1–2
Clams, fried	20–24
Cod fillets, raw, breaded	0–2
Cod fillets, fried	5–10
Fish sticks, fried, frozen	10–15

Chicken

In addition to the following, a review of the basic principles of deep frying will be helpful.

Fresh Chicken

Fresh chicken will produce the highest-quality fried chicken, with the best flavor, eating quality, and moistness.

Chicken Selection

USDA Grade A Broilers that are dressed and drawn and $1\frac{3}{4}$–$2\frac{1}{2}$ lbs in size should be used. The proper chicken size for open kettle frying is $1\frac{3}{4}$–2 lbs. A good chicken size for pressure frying is 2–$2\frac{1}{2}$ lbs. Chicken can be purchased whole or precut into 8, 9, or 10 pieces per chicken. The 8-piece per chicken cut is typically used in open kettle frying. The 9-piece and 10-piece per chicken cuts are typically used in pressure frying and are normally used by fast-food chicken operations. The 10-piece per chicken cut is used by some fast-food operations that are not predominantly chicken specialty houses. The difference in the number of pieces in these cuts is due to the breast being cut into two, three, or four sections.

When whole chickens are purchased, they should not be cut more than 24 h before frying to ensure the highest quality in the finished fried chicken (flavor, eating quality, and moistness).

If precut chickens are purchased, they should be fried within 24 h. This makes daily deliveries a necessity to obtain the highest-quality finished fried chicken.

It is recommended that fresh chicken not be stored for more than 2 days to minimize the possibility of bacterial problems.

With foods, bacterial problems can be of two types. The more severe type is the presence of pathogenic microorganisms, those which are disease producing. With poultry, it is salmonella. The resulting infection, salmonellasis, may kill the very young and very old. Most infected individuals wish they would die. Salmonella, as of this writing, again tops the list of organisms causing food-borne illnesses from all sources. Listeria is currently in second place. Shigella is of considerable concern because it can cause fatalities in individuals with immunological problems.

Obviously, a contamination problem impacting consumers is a concern to the Food and Drug Administration. Brand franchises have been lost as a result of the manufacturing plant environment being contaminated with salmonella.

Microorganisms can also result in off-quality finished product, primarily poor flavor. An example is pseudomonas. This class of organism is commonly found in a wet plant environment such as found in a poultry plant.

Select a good purveyor of fresh chicken. Processing plants must conform with sanitary regulations. Chicken is processed under federal or state meat inspection at all times. The plants scale and separate birds according to weight. Each crate of chicken is packed with ice throughout and on top of the chickens, or it is deep chilled or chill packed by blasting chickens with liquid nitrogen. Delivery trucks are all refrigerated.

Handling of Fresh Chicken by User

Upon receipt of the chicken, weigh it to make sure it is the correct size and weight. Drain water from melted ice immediately and wipe up all spills. Puddles formed by the melted ice water provide ideal areas for extensive microbial growth. The draining of the water from the melted ice prevents the possible waterlogging of the chicken, which leads to quality deterioration.

Place drained crates in a walk-in refrigerator at 32°F (0°C). Crates must be stored off the floor on metal risers, which will permit draining. Floors, walls, and ceilings of walk-in boxes must be cleaned thoroughly on a regular basis with an industrial detergent containing a sanitizer such as chlorine to prevent microbial growth of psychrophilic organisms. These are organisms that grow at low temperatures; they are responsible for most off flavors in fried chicken, which can be described as fruity, musty, fishy, and/or metallic.

Chicken should be stored under refrigeration throughout the complete operation. If an egg and milk dip is used, this should also be refrigerated to minimize bacterial growth. All egg and milk dip must be thrown out at the end of the day.

Storage of Cut Chicken

Place five cut-up chickens in a plastic container with a lid or in a polyfilm bag that can be sealed. For best quality, store this meat in a walk-in refrigerator at 32°F (0°C) for a maximum of 24 h before cooking. Never store chicken for more than 2 days because of the danger of bacterial growth.

Frozen Chicken

Raw, frozen chicken can be purchased for convenience. New freezing methods have greatly reduced the difference between fresh and frozen chicken. When selecting a purveyor, follow the same rules as for fresh chicken. The processor should use quick-blast or nitrogen freezing and make deliveries in freezer trucks.

Thaw frozen chickens before breading and frying. Never fry chickens frozen, as this may cause the blood trapped in the marrow of the bones to turn the bones black. Never refreeze thawed frozen chicken, both for bacterial reasons and because the flavor, eating quality, and moistness will be impaired by refreezing.

Frozen, raw, and prebreaded chicken is available that has been marinated before prebreading and freezing. (This is normally the 10-piece/chicken cut.) This marination step permits the chicken to be fried frozen without the bones turning black. The small pieces obtained with the 10-piece cut permit quicker frying from the frozen state.

Frozen, precooked, and prebreaded chickens are usually small ($1\frac{1}{4}$–2 lbs). The small size and the precooking greatly reduces the fry time required in the restaurant kitchen. Frozen, precooked, and prebreaded chickens should be fried direct from the freezer.

Egg and Milk Dips

Egg and milk dips are used for all kettle-fried chicken. Egg and milk dips can be used for some pressure-fried chicken; however, there are breading mixes that are formulated for pressure-fried chicken that do not require an egg and milk dip. Breading mixes greatly reduce labor and time required for the breading process.

Egg and milk dips are usually more important for open kettle frying than pressure frying due to the fact that with open kettle frying the moisture escapes more rapidly from the chicken, sometimes causing the breading to be blown off the chicken. With pressure frying, the boiling point of water is raised and the breading is not boiled off or blown off as rapidly. Therefore, breading adherence is not as much of a problem in pressure frying.

Formulas for Egg and Milk Dip.

INGREDIENTS	FORMULA 1	FORMULA 2
Large whole eggs	3 doz.	3 doz.
Milk solids	$1\frac{1}{2}$ lbs	1 lb
Cultured buttermilk	–	$\frac{1}{2}$ gal
Water	$1\frac{1}{2}$ gal	1 gal

These formulas should be used for open kettle-fried chicken.

Chicken should be placed into the seasoned flour breading first, then into the egg and milk dip, then back into the final breading or seasoned flour breading. If a thicker coating is desired, the last two steps may be repeated.

If the chicken is to be pressure fried, the chicken may be placed directly into the egg and milk dip, then into the final breading or seasoned flour breading. In pressure frying, it is not as important to seal the pores of the chicken to keep the moisture in.

The use of a wire mesh basket to lower the chicken pieces into the egg and milk dip and then to drain the chicken pieces greatly reduces the time required for this step.

Coatings

Breading.

Breadings used today are generally highly seasoned. The level of seasoning is lower when the chicken is to be open kettle fried; this is because the open kettle-fried chicken has a thicker coating of breading than chicken to be pressure fried. Many good prepared breading mixes are available.

Batter.

If a batter is to be used, submerge the chicken in the batter. Remove the chicken from the batter, allowing it to drain, and place it in the fryer one piece at a time.

Frying Procedures

Review procedures outlined earlier in this chapter.

Sanitation

Recognizing Bacteria Problems in Chicken

Off flavors in chicken, described as fruity, sour or musty flavors in the cooked meat, suggest the possibility of microbial change such as that produced by a type of organism of the pseudomonad group.

As poultry is normally kept refrigerated or even frozen, psychrophilic organisms [organisms producing optimum growth at low temperatures—40°F (4.4°C) and below] are often responsible for the problem.

The source of psychrophilic contamination can be cold storage boxes, loading platforms, and drain water from chicken if it is permitted to accumulate in puddles on the floor. Areas where chicken crates are placed after emptying are prime sources of contamination. The melting ice keeps floors wet, and drippings from the chickens supply nutrients that promote extensive microbial growth in these puddles.

The elimination of these causes can only be accomplished through scrupulous sanitary practices. Recognition of gross microbial contamination need not always be dependent on laboratory analyses. Any of the following conditions could indicate potential trouble areas:

Physical cleanliness: In general, microbial populations are related to the amount of soil present. Ordinary garden soil generally contains from 1 to 10 million microorganisms/g. If water and nutrient material are present, this population can grow to proportions in excess of several billion organisms/g of soil.

Odors: Abnormal odors should be noted. It is often possible even to identify the types of organisms present by their characteristic odor.

Wet areas: When nutrient material is available, wet or damp areas serve as exceptionally fine breeding grounds for microorganisms. Dryness inhibits microbial growth and reproduction.

Slime: The presence of slime is itself visible evidence of microbial growth. As little as a gram of such material will contain in excess of 100 billion organisms. A common but generally overlooked place where such contamination can be found is around refrigerator drains.

People: A most important factor is personnel. The program can be no better than the people who implement it. The most rewarding investment that can be made is in the selection, training, and supervision of personnel as concerns proper sanitation.

Control of Microbes.

Microbials are generally referred to as being of either the "-static" or the "-cidal" type. The suffix "-static" (as in microbistatic and bacteriostatic) refers to control by limiting the numbers of organisms without necessarily killing them all. Chemical agents that accomplish this generally do so by inhibiting the growth of cells. The suffix "-cidal," on the other hand, indicates that the agent kills the organisms. Many antimicrobial agents are "microbistatic" at low concentrations but "microbicidal" at high concentrations. The preferred method is to kill the microbe. Chemical germicides usually fall into one of the following categories:

Halogens: chlorine, bromine, fluorine, and iodine

Phenolics: cresols, phenol, or combined phenols

Quaternary ammonium compounds

Aldehydes, ketones, and alcohols

Bis-phenols and other miscellaneous chemicals

The most effective agents available are the chlorine types.

A good sanitation program can be either an offensive (proactive) or a defensive (reactive) measure. Used offensively, it can result in better-quality products with corresponding economic benefits. If used only defensively, it becomes simply another costly part of the overhead. Obviously, an excellent sanitation program must be a part of a quality assurance program in every commercial poultry processing plant.

Fried Onion Rings

Selection of Onions

Many varieties of onions are used, but the larger types are generally preferred. The more important considerations are to be sure that the onions are hard, clean, bright, have dry skins, and show no sign of decay. Onions will, of course, vary greatly in price, size, and flavor, according to the season.

There are some processors who specialize in cleaning and cutting onion rings for sale to both the foodservice trade and commercial frying operations.

Preparation of Onions

Onions should be washed and peeled with care to remove only the outer, paperlike covering.

Breading of Onions

Onion Ring Coating

INGREDIENTS	lb	oz	METHOD
All-purpose flour	2		Mix flour, meal, and salt together
Cracker meal (medium)	5		
Salt		4	
Egg–milk batter (4–6 whole eggs to 1 pt. milk)			Beat eggs and milk together

1. Roll rings in an all-purpose flour.
2. Dip rings into the egg–milk batter.
3. Coat with the above mixture of flour, cracker meal, and salt.
4. Repeat steps 2 and 3 until the desired amount of coating adheres. Usually onion rings are best when double dipped.

Alternate Breading Method

Some prepared breading mixes also give excellent results. Select a good-quality product and follow the directions of the manufacturer.

Frying: Deep fry for 4 min at 350°F (177°C).

Production hints: Do not handle onion rings any more than necessary after they have been coated. One method of preparing onion rings with a minimum amount of handling is described below.

1. The onion rings are coated in the usual way and weighed in individual-portion paper boats.
2. When an order is received, the individual serving is slid off into the frying basket without handling.
3. The onion rings are fried as usual.
4. The frying basket is drained and the onions are slid back into the same paper boat. They are then served in the same container while they are still hot.

Good-quality onion rings are produced commercially using continuous frying operations. Good-quality onions, good equipment, proper onion preparation, and good frying procedures will result in good end result.

SUMMARY

Deep-frying is a very important usage for oils and fats whether in large food processing plants or relatively small fast-food outlets. This chapter covered the mechanism of the deep-frying process and what happens to the frying oil and the foods that are fried. The basic principles of good frying practice apply to both large manufacturing operations as well as small unit outlets. Where differences existed between some large automatic processing equipment and small batch equipment, these differences were addressed. How to select frying oil and frying equipment, how to take care of both, and the proper preparation and handling of foods, along with some quantity formulas or recipes were also included.

References

1. Swern, D. 1979. *Bailey's Industrial Oil and Fat Products*, 3rd ed. New York: Wiley-Interscience, p. 372.

2. Brooks, D. 1991. Inform, 2(12) 1091.

3. Applewhite, T. 1985. *Bailey's Industrial Oil and Fat Products, Vol. 3*, New York: Wiley-Interscience p. 114.

4. Applewhite, T. 1985. *Bailey's Industrial Oil and Fat Products, Vol. 3*, New York: Wiley-Interscience, p. 114.

5. Lawson, H. 1970. *Fats and Oils for Poultry Processors*. Chicago, IL: Institute of American Poultry Seminar on Quality Standards.

General References

Varela, G., Bender, A., and Morton, I. 1988. *Frying of Food: Principles, Changes, New Approaches*. Chichester, England: Harwood.

Chapter 8

Griddling and Panfrying

The use of oils and fats for griddling and panfrying is almost exclusively a foodservice practice. However, in order to provide a rather complete coverage of all of the major uses for food oils and fats, this subject matter should be included.

Griddling and panfrying are the major methods of food preparation for breakfast items. About 75–80% of all breakfast menu items are prepared on the griddle or in the fry pan. In recent years, many fast-food chains have increased sales volume and profits by adding a breakfast menu. This has enabled them to make more efficient use of overhead costs and reduce some fixed unit costs. In addition, a successful breakfast operation can help build an even better lunch and dinner business.

GRIDDLING TECHNIQUE

There are four basic steps to a successful griddling operation:

1. Always start with a clean griddle and clean it regularly.
2. Season the griddle properly with the griddling oil every day.
3. Grill foods at the right temperature and recheck the temperature regularly. A common mistake is to use too high a temperature for many griddled foods.

4. Use the best griddling oil or fat available and in the right amount. The use of too much griddling oil or shortening is a very common mistake.

Care and Cleaning of the Griddle

It is important to keep the griddle clean at all times, for three principal reasons:

1. Foods look more appealing and taste better when they are prepared on a clean griddle, as compared with one with burned food particles and excess grease that gets picked up when the next foods are fried.
2. The griddle is easier to maintain. Regular cleaning will minimize the need for periodic backbreaking work to scour off the grease, gum, and burned food buildup.
3. It reflects on the overall image of the establishment. A dirty griddle in view of customers gives them an image of sloppy, unsanitary practices throughout the operation and a bad impression about the people who work in the operation.

Maintaining a clean griddle avoids messy, time-consuming, major cleanup jobs.

Various degrees of cleaning can be employed in a griddle operation. They can be broken down into three types of cleaning, and they should be employed every day; they are continuous, intermediate, and total cleaning.

Continuous Cleaning

Use a scraper or spatula to completely scrape the griddle surface after each food order is prepared. This removes burned food particles and prevents an excessive buildup of griddling fat and gum.

Intermediate Cleaning

During slack periods, pour 2–3 oz of water on the hot griddle. Heat the griddle until it sizzles and then scrape. This loosens particles that were missed in continuous cleaning. Note: Wear gloves and

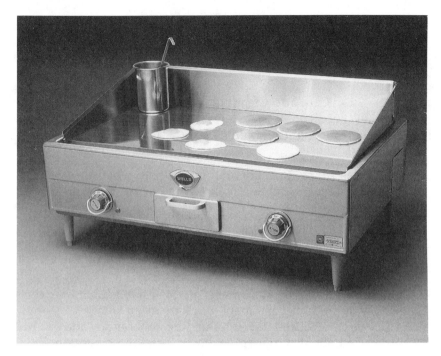

FIGURE 8.1. Foodservice griddle. (Courtesy of Wells Manufacturing Company.)

avoid spattering, which can burn the skin. Do not allow water to dry on the griddle surface, because this may require reseasoning the griddle.

Total Cleaning

At the end of the day, clean the griddle surface down to bright metal by using a pumice stone, a griddle screen, or a carborundum griddle stone. This removes all food particles, grease, gum, and seasoning from the surface. Use a block of soft carborundum where available, as it is relatively low in cost. Care should be taken to avoid scratching the griddle. A deep scratch can trap added gum and cause food to stick in the future.

After the total cleaning, the griddle must be properly seasoned before it can be used again.

Seasoning the Griddle Surface

A new griddle or newly cleaned griddle surface must be properly seasoned in order to do a good griddling job. Seasoning is the building of even layers of griddle oil or shortening until a slick, hard surface is formed. Stop at this point, where a minimum of food sticking will develop, but not to the point where it becomes a solid varnish-like in appearance.

The seasoning is carried out in the following manner. The griddle is first heated to 325°F (163°C), after which a very small amount of griddling oil or shortening is applied. This small amount is then spread with a cloth to absorb the excess and create a thin oily film on the griddle surface. This complete procedure is repeated several times until a slick, firm surface is created. The condition of the griddle will determine the number of applications necessary for good release of food.

Temperature Control

Frying at the right temperature is one of the more important factors in preparing better-looking, better-tasting griddled foods. Food prepared at too high a temperature may also be undercooked inside, even though the outside appears done.

Most griddles in use today have different temperatures in different locations across the surface because of different shapes and locations of their heating sources. Before preparing any food on the griddle, it is important to identify where the "hot spots" and "cold spots" are so that each food can be fried at its proper temperature. A special surface thermometer can be used to give good guidelines on which to operate.

Proper Griddle Temperatures

	°F	°C
Eggs	300–325	149–163
Bacon	325	163
Sausage	350	177
Hash browns	350	177
Grilled sandwiches	350	177
Chops	350	177

Cutlets	350	177
Steaks	350	177
Hamburgers	350	177
Pancakes	400	204

Griddle Shortening

Use the best griddling shortening available, and use the right amount. Some of the more important types of products used for griddling are butter, margarine, cooking/salad oil, and specialty griddling shortenings.

When butter is used, it is generally to promote the relatively high image rating normally associated with butter. Good results can be obtained by using a good-quality butter and by adhering carefully to good griddling practices. Good, fresh butter will generally result in good-flavored griddled foods. The disadvantages of butter are as follows: (1) It is relatively costly, (2) it is less convenient to use, and (3) it can give griddled foods a poor appearance. Butter is generally expensive in relation to other griddling fats, and about 25% of the butter is usually wasted. When used for griddling, butter is generally clarified by melting and removing the fraction containing water, milk solids, and salt. If butter is not clarified, the milk solids will tend to contribute to scorching of the griddled foods. Even when clarified, butter does not contain antisticking agents or antigumming agents, and care must be taken especially to prevent food from sticking to the griddle. Eggs require careful griddling.

Margarine products are used successfully in some operations. Most margarines require the same precautions as butter. The margarine must be of good flavor and texture, and care must be used to obtain good griddled food products. Milk solids from margarine can cause the same scorching problems as butter. Low-temperature griddling is imperative when using butter or margarine. Margarine is also clarified, resulting in about the same 25% average waste as butter. In addition, the buttery flavor of margarine tends to be distilled off under the heat of the griddle.

Salad oils and cooking oils are easy to use and lend themselves well to controlling the quantity to be used. Their disadvantages include the following: They tend to gum on the griddle surfaces, especially under high-temperature use; they do not have a buttery flavor and sometimes impart rancid flavors to griddled foods; and they generally do not contain antisticking agents.

Specialty griddling shortenings have become the favorite products for lubricating the griddle. Two classes of products are in general use: (1) products that are solid at room temperature, usually containing from 25 to 50% coconut oil, and (2) fluid vegetable shortenings that are liquid at room temperature, generally formulated from partially hydrogenated soybean oil.

The solid-shortening-type griddling shortenings contain an antisticking agent (usually lecithin) as well as a yellow color and an added buttery flavor. These products as a group do a good mechanical job on the griddle and are generally more satisfactory for an additional use, that of lubricating bun toasting machines. In the latter, where ultrahigh temperatures are employed, coconut oil provides good resistance to gumming. Products containing coconut oil have, of course, suffered a loss in image rating due to their high level of saturated fatty acids. This was discussed in greater detail in chapter 5.

Liquid- or fluid-type griddling shortenings contain ingredients to prevent sticking, minimize gumming, and provide a golden yellow color and buttery flavor. Their liquid form offers a great opportunity for ease of use and cost control. Portion-control squeeze bottles can be used without having to melt the product.

Another type of specialty griddling shortening that has come to market is an offshoot of a soybean-based fluid shortening. In addition to containing a buttery flavor and antisticking and antigumming agents, it contains salt to provide additional buttery flavor impact.

PANFRYING TECHNIQUE

Most panfrying is done with about $\frac{1}{2}$-inch of butter, margarine, or specialty griddle shortening in the fry pan, especially for foods such as chicken, fish, and other dinner entrees. However, the panfrying of eggs and other breakfast items is generally different. The fry pan should be seasoned in a manner similar to that used for the griddle surface. The fry pan should be warm (about 325°F; 163°C) and a small amount of shortening should be poured into the pan. Then, using a cloth, the pan surface should be rubbed until a mirrorlike finish is obtained. Once the pan is seasoned, it should be wiped clean and not washed. This seasoning permits eggs to be panfried with a minimum amount of oil or shortening and produces finished foods that are appetizing.

Typical panfried menu items include (1) meats and seafoods such as steaks, chops, hamburgers, chicken, and fish (whole or fillet), (2) breakfast items such as eggs, omelets, sausage, ham slices, and bacon; and (3) vegetables such as potatoes and eggplant.

Preparation of the Skillet for Pan Frying of Eggs

The fry pan is usually held at low heat, ready to use. Specialty griddling shortening is poured into the skillet and then swirled to provide an even coating. The excess shortening is then poured back into the container ready for the next usage. The thin coating permits the cook to produce eggs with excellent eye appeal and flavor.

DO'S AND DON'TS FOR A GOOD PANFRYING OPERATION

1. Do not wash the frying pan; only wipe it clean.
2. Do not use excessive amounts of shortening; merely coat the skillet and return the excess to the container.
3. Do not fry at too high a temperature; usually a low heat (about 300–325°F; 149–163°C) is most satisfactory for panfrying eggs.

SUMMARY

This is a brief coverage of the use of oils and fats for griddling and panfrying, which are primarily foodservice applications. It included the advantages and disadvantages of various types of oils and fats, the proper preparation of foods, and care and cleaning of equipment.

Chapter 9

Salad/Cooking Oil Usage

COOKING OIL USAGE

Cooking oils are a type of liquid oil, generally of vegetable origin, used in food processing plants, foodservice operations, and in the home. Marine oils are the only cooking oils derived from animals. Natural marine oils are highly unsaturated, which results in products with poor flavor stability. Hydrogenated marine oils are used in cooking in some parts of the world.

Cooking oils are used either in their natural state or after processing, depending on such factors as local taste, cost goals, and the flavor of the source oil. For example, olive oil is used in Mediterranean countries because it is readily available at a reasonable cost, and the people like its natural flavor. On the other hand, cottonseed oil contains objectionable flavors and odors as well as very high levels of color bodies. Therefore, cottonseed oil must be refined, bleached, and deodorized before use as a cooking oil (1).

In addition to olive oil, there is a substantial market for "natural" peanut, sesame, and safflower oils as cooking oils in many European, Asian, and South American countries. In the United States, use of such oils is becoming more popular in recent years because of nutritional considerations.

SALAD OIL USAGE

There is a significant difference between cooking and salad oils. The term salad oil is generally reserved for those products that remain substantially liquid at refrigerator temperatures (about 40°F or 4.4°C). Sometimes this is referred to as resisting graining at refrigerator temperatures.

The primary use for salad oils is for making salad dressings. Other uses include deep-frying, panfrying, griddling, and some types of baking. Salad oil may be used as a cooking oil, but cooking oil cannot be used as a salad oil (except for olive oil).

It should not be assumed, however, that salad oil can be used for deep-frying, panfrying, griddling, or baking with the same success and standards of quality as the oil or fat products specifically designed for those uses.

Salad oil is a clear, light-colored vegetable oil. It can be poured, metered, and pumped. Salad oils are made from canola, olive, cottonseed, sunflower, peanut, safflower, and from partially hydrogenated soybean oil (2). These oils are found in seeds or seedlike parts of these plants. They have varying degrees of flavor and oxidative stability, depending on the base oil and manufacturing standards. Most oils are refined, bleached, and deodorized to remove flavor and are sometimes lightly hydrogenated. Olive oil, whether used as a salad oil or as a cooking oil, is generally not deodorized because it lends a special flavor to salads and Italian foods that is generally considered to be desirable.

Salad oils utilizing cottonseed oils are generally "winterized" to remove those small portions of the oil that would solidify at refrigerator temperatures (3). The addition of very small quantities of crystal inhibitors, such as polyglycerol esters, provides further protection against graining at low temperatures. A small amount of graining or oil crystallization can accelerate mayonnaise or salad dressing separation at storage temperatures.

Most salad dressings depend on the oil mainly to coat salad ingredients and to carry the flavors of herbs, spices, and vinegars.

Soybean salad oils are the most prominent types today because of the availability and value of soybean oil and the use of the latest improvements in processing techniques. Crude soybean oil is refined, bleached, lightly hydrogenated, and sometimes winterized to remove the small amount of higher-melting fat fractions. Then it is deodorized to remove odors and flavors. Finally, materials such as polyglycerol esters and methylsilicone may be added, and the product is packed. When one or more of these materials are added to salad oil, it is usually to improve panfrying, griddling, or deep-frying performance. However, the latest trend appears to be to use smaller amounts or no additives at all. This is because the no-additive product sometimes has enhanced marketing or promotional opportunities that might outweigh the performance of these additives.

The use of soybean oil without hydrogenation is growing in importance for both cost-saving and possible nutrition claims.

As mentioned previously, plant breeding programs have resulted in new oils such as canola, high oleic safflower, and high oleic sunflower oil that are suitable as salad and cooking oils (4).

When canola, sunflower, or olive oils are used alone or in combination with other oils it is generally to promote nutritional or health claims; for example, low in saturated fat.

Sunflower seed, safflower seed, canola, and corn oils may need dewaxing before they can qualify as a salad oil.

When salad oil is used in baking, it is generally used to make chiffon-type cakes and in some all-purpose baking. Examples of the latter products are biscuits, cornbread, and dinner rolls.

Traditional salad dressings, whether made by large food processors or made in small batches in foodservice establishments, are relatively high in oil content. They can be as high as 80% oil in many dressings. Many products have appeared on the market with claims of being low in oil/no oil, low cholesterol/no cholesterol, and so on. Some that are actually low in oil are very poor in quality. Some are obviously mislabeled. Some labeling states an unrealistic serving size such as one tablespoon or less. Then if the dressing has some oil and/or cholesterol in the product, by cleverly rounding off the figures they are allowed to make a claim of no fat or oil/no cholesterol. In the future, everything points to the Food and Drug Administration taking a much harder line in policing irrational claims.

In addition to policing their own nutritional claims, the food processor should be careful to follow the rules where Federal Standards of Identity apply. These regulations state that to label products as mayonnaise, salad dressing, or French dressing the minimum amount of vegetable oil must be 65%, 30%, and 35%, respectively. If any of these contain less than the minimum oil, they should have different names to identify them.

One result of the current interest in body weight control and nutrition has been the introduction of reduced-calorie food products. To qualify as "reduced calorie", the product must contain one-third fewer calories than the conventional product. In the case of dressings, the reduced-calorie product must contain less oil and/or carbohydrate and more water than conventional dressings.

One rule of thumb to employ in making "low-calorie" dressings is to merely replace one-third to one-half of the oil in the quantity recipe or formula with something else. The replacement may consist of one or more of the following: lemon juice, vinegar, tomato juice, buttermilk, prepared vegetable juice, fruit juices, or low-fat yogurt. When making significant replacements such as these, the use of fiber, such as a food cellulose, may be needed to obtain the proper consistency. Considerable experimentation may be required to achieve the desired flavor and consistency (5).

MAJOR TYPES OF SALAD DRESSINGS

There are three basic categories of dressings used to enhance salads: (1) the mayonnaise-type dressings, (2) salad dressings with some type of cooked starch paste, and (3) French-type dressings of both the separating and non-separating variety.

Mayonnaise-Type Dressings

Basically, mayonnaise types are emulsions of oil in water. Ingredients such as eggs and edible gums are used to stabilize or thicken the emulsions; other ingredients are added for flavor, color, and piquancy. The stability of the emulsion depends on ingredients (especially an oil that resists graining at refrigerator temperatures), the equipment, and how it is operated. In large plant operations, specialized equipment such as colloid mills and intense mixing systems are used in order to reduce the oil to a very low particle size. Lower processing temperatures (40–50°F; 4.4–10.0°C) may also contribute to emulsion stability.

Salad Dressings

This group differs from the mayonnaise type by using a lower level of salad oil (30–40%) and by employing cooked starch materials, emulsifiers, and gums to provide stability and thickness. These dressings are similar in appearance to mayonnaise, somewhat similar in taste, and a cheaper product; sometimes they are confused with mayonnaise.

French Dressings

A separating French dressing is a temporary emulsion of oil, vinegar, and/or lemon juice, and seasons. These temporary emulsions usually separate shortly after mixing and require shaking before each use.

To make nonseparating French dressings, ingredients such as egg yolk and/or other emulsifying ingredients are used to keep the oil in suspension. In addition many stabilizing gums can be used for even further emulsion stability. These include gum acacia, gum arabic, carob bean gum (locust bean gum), gum tragacanth, Irish moss, cellulose gum, xantham gum, and so on.

According to the Federal Standard of Identity, the minimum amount of vegetable oil that a mayonnaise can contain is 65%. Many, however, contain a higher percentage, sometimes as high as 75–80%. Naturally, then, it is of vital importance to use a good oil—one that is bland and delicate in flavor, is uniform in quality, and leaves no unpleasant, oily aftertaste. Usually, as the percentage of oil is decreased, the quantity of egg yolks or whole eggs necessary to make a good-bodied mayonnaise increases. This makes the challenge of good-quality low-oil/no-oil/low-cholesterol/no-cholesterol dressings a difficult one.

FUNCTIONS OF INGREDIENTS FOR SALAD DRESSINGS

Eggs

Eggs are extremely susceptible to spoilage. Care must be taken in their use; bacterial count increases rapidly if they are exposed to air at room temperature after being broken, thawed, or washed.

Vinegar

Vinegar is responsible for much of the flavor of dressings and also helps to give it good keeping qualities during storage.

Cider vinegar and white distilled vinegars are the two types most often used. Cider vinegar makes a mayonnaise or salad dressing with a less sharp taste and with a better "bouquet" or odor. The strength of vinegar is expressed in terms of "grain," which is 10 times its acidity; for example, a vinegar containing 5% acetic acid is known as 50-gr vinegar.

For a distinctive flavor, lemon juice is often substituted for part of the vinegar.

Salt

Salt is an important ingredient. It is a flavor enhancer that helps to bring out the flavor of the other ingredients. Salt helps to preserve the dressing, giving it stability and body. The usual range of salt in commercial mayonnaise is 1.2–1.5%.

Spices

Mustard provides added flavor and a small amount of color. The amount of dry mustard used varies with the strength of the mustard and the type of flavor desired. The usual range is 0.1–1.0%. Mustard flour is used to a considerable degree. Both white, which is hot to the taste, and brown, which has a sharp odor, are generally blended to obtain the desired level of flavor and pungency.

Pepper is used for its flavor. If used, it should be a good grade of white pepper, free of any black specks (decorticated white pepper).

Paprika is used primarily for color. It adds very little flavor. (The Hungarian variety has more flavor than the Spanish.)

QUANTITY FORMULAS FOR SALAD DRESSINGS

French Dressing (Separating Type)

Ingredients	1 gal Oil Batch	Method
Paprika	3/4 cup	Mix first 7 ingredients
Mustard	4 tbsp	together to form paste
Salt	3/4 cup	

Catsup	1 cup	
Sugar	2 cups	
Garlic (minced) or onion	To taste	
Water	1 cup	
Salad oil	1 gal	Whip in about one-quarter of the oil. Continue whipping and add remaining oil and vinegar alternately.
Vinegar (cider)	1 qt.	

Mayonnaise-Type Dressing

Ingredients	Large Batch Measure	Method
Whole eggs[a]	35–40 eggs	Beat at high speed of mixer for 2 min.
Salt	1 cup	Mix dry ingredients together.
Sugar	1 cup	Add to beaten eggs and
Mustard (dry)	1 cup	beat 2 min at high speed.
Salad oil[a]	5 gal	Add a fine stream, gradually increasing flow of oil as amount of emulsion builds up. Add all oil in 10–15 min at high speed.
Vinegar (50 gr, cider)	$3\frac{1}{2}$ pt.	Add small amounts as needed during addition of oil to keep emulsion from breaking. Add remainder after all oil is in, with mixer running at low speed. Scrape down bowl and mix for an additional minute.

Yield: Approx. 56 pt.

[a]For best results, have eggs, salad oil, and vinegar cold (50–55°F; 10–13°C).

Tartar Sauce

Ingredients	Large Batch Measure	Method
Mayonnaise-type dressing	1 gal	Blend to a uniform mixture
Pickle relish (drained)[a]	3 pt.	
Parsley (chopped)	$\frac{1}{2}$ pt.	
Onions (chopped)	1 pt.	

Yield: Approx. 12 pt.
[a]Chopped dill pickle or a sour pickle may be substituted if desired.

Thousand Island Dressing

Ingredients	Large Batch Measure	Method
Mayonnaise-type dressing	1 gal	Blend to a uniform mixture
Chili sauce	1 qt.	
Pickle relish (drained)	1 qt.	
Pimiento (chopped)	$\frac{1}{2}$ pt.	
Paprika	$1\frac{1}{3}$ tbsp.	

Yield: Approx. 12 pt.

Creole Remoulade Dressing

Ingredients	Measure	Method
Brown creole or Trieste mustard	$\frac{1}{2}$ cup	Mix all ingredients thoroughly and pour over 8
Tarragon vinegar	1 pt.	doz. boiled shrimp. Let
Paprika	6 tbsp.	marinate together for at
Cayenne pepper	$1\frac{1}{2}$ tsp.	least 4 h. Serve cold on
Garlic powder	$1\frac{1}{2}$ tsp.	shredded lettuce.
Catsup	$\frac{1}{4}$ cup	
Salad oil	1 qt.	
Green onions and tops, finely chopped	1 cup	
Celery, finely chopped	1 cup	
Salt	1 tsp.	

Note: If a thick, smooth consistency is desired, substitute 1 qt. of mayonnaise-type dressing in place of 1 qt. of salad oil.
Yield: Approx. 4 lb, 4 oz ($1\frac{1}{2}$ qt.).

Coleslaw Dressing

Ingredients	Measure	Method
Malt vinegar	1 qt.	Blend together the salt,
Salt	$\frac{1}{4}$ cup	sugar, paprika, pepper,
Light brown sugar	$\frac{1}{2}$ cup	oregano, parsley, celery
Paprika	2 tbsp.	seeds, and pimiento with
Black pepper	1 tbsp.	the vinegar until the salt
Chopped fresh parsley	1 cup	and sugar are dissolved.
Oregano	2 tbsp.	
Celery seeds	4 tsp.	
Pimiento or red sweet pepper, chopped	$\frac{1}{2}$ cup	
Salad oil	3 qt.	Add the salad oil to the above and blend thoroughly. Lightly coat finely shredded cabbage and marinate in refrigerator overnight.

Yield: Approx. 1 gal.

SUMMARY

This chapter covered cooking and salad oils including the source oils used, their differences, and processing steps employed. Also directions for using salad oils in making mayonnaise and other dressing types were discussed.

References

1. Latta, S. 1991. Inform, 2(2), 105.
2. Institute of Shortening Edible Oils. 1988. *Food Fats and Oils*, 6th ed. Washington, DC: *Institute of Shortening and Edible Oils*, p. 20.
3. Swern, D. 1979. *Bailey's Industrial Oil and Fat Products*, 3rd ed. New York: Wiley-Interscience, p. 256.
4. Institute of Shortening Edible Oils. 1988. *Food Fats and Oils*, 6th ed. Washington, DC: Institute of Shortening and Edible Oils, p. 20.
5. Marsili, R. 1993. Food Product Design, 49–50.

General Reference

Swern, D. 1982. *Bailey's Industrial Oil and Fat Products*, Vol. 2, 4th ed. New York: Wiley-Interscience, pp. 315–337.

Chapter 10

Baking Technology (Including Icings for Baked Goods)

This chapter covers baking technology, including cakes, icings, pies, sweet yeast goods, and cookies. The basic principles apply to any baking operation, be it wholesale baking, retail baking, foodservice commissary, in-store supermarket baking, and so on. Quantity recipes/formulas must be adjusted in batch size to meet the specific needs of a particular operation. Obviously, the larger the operation, the greater the need for larger and more automated equipment.

CAKE BAKING: GENERAL

In foodservice cake baking, prepared baking mixes are used to a considerable extent. Shortenings used in these mixes are generally partially hydrogenated vegetable oils and include edible emulsifiers, such as monoglycerides and diglycerides. Others include some more sophisticated emulsifiers, such as lactic acid esters or propylene glycol esters of fatty acids.

When cakes are not made from a mix, either a plastic shortening or a fluid shortening may be employed. The plastic shortening is generally packed in 50-lb cartons and contains monoglyceride and diglyceride emulsifiers. The fluid cake shortening generally contains one of the newer, more sophisticated food emulsifiers. The reason for this is that plastic shortenings provide a certain amount of creaming or air incorporation owing to the nature of their plastic form. Fluid shortenings must depend almost entirely on the emulsifier system for air incorporation and, therefore, require more potent emulsifiers.

Plastic monoglyceride-type cake shortenings (high-ratio shortenings) permit the use of relatively high ratios of sugar, shortening, eggs, and water in relation to the flour in the batter. This results in better eating qualities, keeping qualities, texture, and flavor in the finished baked cake than can be obtained with a regular, nonemulsifier shortening.

The use of fluid shortening generally results in even better finished-cake keeping qualities. This is because fluid shortening with the more potent emulsifiers such as propylene glycol monostearate and lactic acid esters makes possible the incorporation of even higher proportions of liquids and holds the liquids for longer periods after baking. It is also possible to use about 10–15% less fluid shortening (by weight) in cake batters than plastic emulsifier shortenings.

FUNCTIONS OF INGREDIENTS IN BAKING

All baking ingredients are classified as one or more of the following: tenderizers, structure builders, moisteners, driers, and flavorers.

Shortening

Shortening, being a fat, is the only major ingredient in the batter or dough that is not changed during the baking process. It is the only ingredient that is not dissolved in water or made wet by water.

In addition, shortening does not combine with other ingredients. It imparts softness and tenderness (1). The dispersion of the fat particles throughout the batter prevents the mix from becoming a hard and tough mass. The degree of shortness or tenderness that a given baked product possesses depends on the type of shortening and the amount of shortening used, as well as on the manner in which the shortening is dispersed in the batter or dough.

The greater part of the aeration is obtained when the shortening is creamed with the sugar or the flour. In products that are not shortened, such as angel food cake, aeration is obtained by whipping the eggs. In layer cakes, or any shortened cake leavened with baking powder, the aeration from mixing augments the leavening action of the baking powder.

The creaming properties of the shortening give the "creamed" icings volume, fluffiness, smooth texture, and the creaminess from which they get their name. Fudge icings would be tough, tacky, or hard if shortening were not used.

In creamed icings, the shortening "creams up light" with the sugar and allows air to be incorporated. As the amount of shortening is increased, the volume of the icing is increased (up to a point). Therefore, higher levels of shortening in icings result in increased volume of cream icings.

During the mixing process, the shortening aids in the development of the structure of the batter or dough by helping to produce tiny air cells or bubbles of microscopic size. The cell structure is influenced by the manner in which the shortening is dispersed in the batter or dough.

The formation of air or carbon dioxide cells is necessary for proper expansion in both leavened and unleavened products. In the cells, the air incorporated in the batter, or the mixture of air and the carbon dioxide from the leavening, becomes saturated with moisture vapor from the liquids in the dough or batter. Then, as the temperature of the batter is raised when the product is placed in the oven, the cells expand and produce the expansion in the finished product.

Because shortening plays an important part in the formation of the cellular structure of the batter or dough, it influences the volume, grain, and texture of the finished product (2).

Shortening's function of contributing to eating quality is related to its primary function of producing shortness and tenderness. Eating quality is a combination of the effects of taste, odor, and physical impressions, such as shortness, tenderness, moistness, dryness, fineness, chewiness, toughness, coarseness, and brittleness.

Flour

Flour is the basic structure-building ingredient in almost all bakery products. Wheat gluten protein along with water forms the major structure-forming material supporting the other ingredients in bakery products. Of the various types of flour, including wheat, corn, and soy, the wheat gluten proteins form the most significant and by far the strongest "gluten" structuring material. A microscopic examination of a cake batter would show tiny air cells encased in a film of fat, which, in turn would be suspended in a "lake" of liquid syrup. A close examination would also show a fine network of fibrous material that, through the process of baking, becomes the structure of the finished product.

The following tabulation shows the approximate composition of typical cake and bread flours:

	CAKE FLOUR (%)	BREAD FLOUR (%)
Moisture	11.0	11.0
Protein	9.0	12.0
Starch	78.0	75.0
Fat	1.7	1.6
Ash	0.3	0.4

Flour Characteristics

Strength

Flour strength means the ability of a flour to carry a heavy load of other ingredients: sugar, shortening, milk, and eggs. This characteristic is generally used only in reference to the wheat flour types.

Flour strength depends on the type wheat and the milling technique. The amount and character of the "gluten" developed by a flour in a dough is usually considered to be the index of the strength of a flour. Strong flours are generally higher in protein content than weak flours. Flour strength also depends upon the source of the wheat. Generally, winter wheats are used to make strong high-protein flours, and spring wheats for lower-protein flours.

Color

Good flour color is important not only in angel food, white pound, and white layer cakes, but in yellow cakes as well. In yellow cakes, the brilliance of the yellow color of the finished cake is enhanced if the flour has a clean white color.

Tolerance

The ability of a flour to make good cakes with some formula variation or mixing abuse is known as "tolerance." Cake flour tolerance is usually related to the amount of sugar or liquids it can carry.

Granulation

Flour granulation refers to the size of the particles of flour. Cake flours are, in general, ground much more finely than bread and pastry flours, because this, in turn, helps to produce cakes that are fine and even-grained along with soft texture.

Absorption

This term refers to the ability of a particular flour to absorb water without producing a slack dough that is sticky and difficult to use, especially in large plants with automatic machinery. It is used most frequently when referring to bread flours.

Acidity

Flour acidity is a characteristic affected chiefly by the bleaching process and storage conditions of the flour. The yellow color of fresh milled flour disappears gradually on aging as a result of oxidation. Bleaching with chemical agents accelerates this process and provides for a greater uniformity of flour color. For this purpose, oxides of bromine, chlorine, or nitrogen may be employed. Cake flours, from the standpoint of highest quality in the finished cake (i.e., color, volume, and keeping qualities), usually have an optimum acidity range (pH 5.0–5.2) that is slightly on the acid side.

Milk

The nutritive value of milk comes from the milk solids, which are made up of butterfat and nonfat milk solids. Whole milk solids contribute fat, protein, carbohydrates, minerals (especially calcium), and vitamins. Nonfat milk solids contribute all of these important elements to the batter except fat. Nonfat milk solids are the most important type of milk product used in the bakery.

The milk solids add their own flavor to that of the other ingredients.

The sugars, principally lactose in the milk solids, partially undergo a browning reaction at baking temperatures, and their presence in the batter or dough improves the color of the crust.

Both whole milk solids and nonfat milk solids combine with water and have a "drying" action, because they remove some of the moisture in the batter or dough, thereby permitting less "bake out."

The moisture portion of liquid milk will act in combination with flour proteins as a structure builder. The water combines with the gliadin and glutenin in the flour to form gluten. The moisture in milk will also improve eating and keeping qualities in high-ratio cake batters. The higher proportion of liquid in relation to flour enables the baker to produce goods with a higher moisture content. These moisture products will retain their initial, good eating qualities over a longer period of time.

Sugar

Sugar performs the following functions:

1. Adds caloric food value.

2. Supports yeast activity for the production of carbon dioxide gas for leavening.

3. Improves keeping quality. High-ratio cakes, with their high proportion of sugar in relation to flour, also require increased proportions of liquids in the batter. Consequently, high-ratio cakes have better keeping qualities. The use of invert sugar, honey, and glucose, all of which have moisture-retaining properties, may further improve keeping qualities.

4. Improves the grain and texture. Products containing ample amounts of sugar have a softer, richer texture and a more uniform grain.

5. Improves flavor.

6. Controls spread in the manufacture of cookies. The coarser the grind of the sugar, the greater the amount of spread during baking. The use of a very fine grind such as 6X powdered sugar will result in a minimum amount of cookie spread.

7. Helps bind an icing together. Sugar is the major ingredient of all icings, representing 50–80% of the weight of an icing. The

quality and texture of an icing depend to a large degree on the type of sugar used. Part of the sugar dissolves in the water that is present and remains dissolved as a syrup. It is this syrup that helps to bind the icing together. A simple water icing that has lost its liquid by evaporation has lost the binding material, and as a result the icing will readily crumble. In addition, crystal size should be small to avoid grittiness to taste. This is especially important for the portion of sugar that does not dissolve in the liquids.

Eggs

Solids in the egg:

1. Contribute color (yolks or whole eggs).
2. Add flavor and richness.
3. Act as tenderizers. Egg yolk and whole egg solids contain fat and natural emulsifiers in the form of phospholipids which contribute tenderness through their shortening action.
4. Function as structure formers. Egg solids, chiefly the egg white solids combined with the moisture in the egg, are considered structure-forming materials that help significantly to produce proper volume, grain, and texture. Egg solids are, therefore, unique because they act both as structure builders or tougheners and as tenderizers (fat in the egg yolk).
5. Add nutritive value. Eggs add high nutritive value to the finished product. They contain important amounts of protein, fat, and minerals (3).

The moisture part of the egg is similar in function to the moisture part of the milk products used in the production of high-ratio cakes. Moisture in the egg will combine with the flour to form gluten, a structure-forming material, and act as a moistener, playing an important part in building keeping qualities.

In addition to the functions performed by the solids and the moisture in eggs, the egg itself may function to leaven or lighten batters for some types of cake, such as sponge and angel food. In these cakes, eggs contribute to lightness and facilitate incorporating and holding air in the batter.

Whole eggs, egg yolks, and egg whites are all used in the bakery. Frozen eggs are used to a greater extent than fresh or shell eggs because the latter are harder to handle and store. In addition, the process of cracking the egg wastes valuable time; there is also considerable waste due to some portion adhering to the shell. Frozen egg products function as satisfactorily as fresh eggs.

Leavening Agents (4)

A leavening agent is a material capable of producing carbon dioxide gas, which inflates the batter or dough to its proper size during the baking operation. This gas supplements the natural expansion of the product from the heat of the oven and is largely responsible for its volume, grain, and texture.

Chemical Leavening Agents

Chemical leavening agents are used when a sweet, non-yeast-flavored product (cake, cookies) is desired.

"Baking powder" is the most widely used term for leavening materials that are combinations of a baking acid and soda. The commercial acids used in baking powder may be cream of tartar, tartaric acid, sodium aluminum sulfate, and monocalcium phosphate, used either separately or in combination.

Most chemical leavening systems are "triggered" by the addition of water (in making batters or doughs), which brings the acids and alkaline materials together to release the leavening gas (carbon dioxide). Because this action starts as soon as mixing begins, the leavening action can be used up before baking starts unless the action is slowed down. This retardation is accomplished both by specially treating the acid particles to slow down their solubility rate and by choosing baking acids that will not react particularly well without the application of heat from the oven to the batter. Heat-activated leavening is particularly useful in batters that must stand for a while before use, for example, pancake batters.

Double-acting baking powders contain both fast-acting and slow-acting acids.

Yeast (5)

Yeast is a fungus, a living organism, that reproduces prolifically and produces carbon dioxide, alcohol, and, to a small extent, flavor compounds under the influence of the proper environment. This environment includes water and yeast food (sugar). Yeast has a mellowing or softening action on the dough structure, and because of this, the dough becomes more easily inflated by the carbon dioxide produced. Yeast is used whenever the product to be produced is to have a fermented or yeasty flavor, for example, in bread and sweet yeast goods.

Cocoa

Cocoa is essentially bitter chocolate liquor obtained from the cocoa bean from which most of the cocoa butter has been removed. To produce the more desirable reddish-brown color and a specific desirable flavor, cocoa is treated with caustic soda; the resulting product is called Dutch process cocoa. Cocoas that have received only a mild treatment of soda are called partially dutched cocoas (6).

CAKE-MAKING TECHNOLOGY

Cake batter is produced by using flour and eggs as structure builders, sugar as a tenderizer, water and/or milk as moisteners, a leavening agent to produce gas, flavoring materials (if desired), and an emulsifier type of shortening to properly disperse the fatty materials in the aqueous phase (7). A finished batter, therefore, is an emulsion of tenderizing and toughening materials in the proper balance to produce cakes of the desired finished texture, volume, and taste.

Mixing

Cake batters can be mixed continuously in specially constructed mixers in large plants, or in individual batches in small and medium size cake-producing plants.

In mixing the cake batter, the essential principle of the blending method is to obtain a smooth batter in the shortest possible time without overmixing. To ensure a smooth batter, scrape down the mixer and the creaming paddle frequently. The batter should be scraped down at least once during each stage of the mixing operation.

Table 10.1. Scaling Weights and Baking Temperature for Cakes

Cake	Pan Size	Scaling weights	Baking Temperatures °F	°C
Sheet cakes	17 × 25 in.	6–7 lbs	360–370	182–188
Layers	8 in. diameter	12–14 oz	375	191
(1$\frac{1}{2}$–2 in. deep)	7 in. diameter	9–11 oz	375	191
	6 in. diameter	7–8 oz	385	196
Loaf cakes	7$\frac{1}{2}$ × 3$\frac{1}{2}$ × 2$\frac{1}{4}$ in.	14 oz	375	191
Ring cakes	6 in. diameter	14 oz	375	191
Cupcakes	2 in. diameter	15 oz/doz.	385–400	196–204
Pound cake	6 × 11 × 3$\frac{1}{4}$ in.	2$\frac{1}{4}$ lbs	300–330	149–166
Pound cake	7$\frac{1}{2}$ × 3$\frac{1}{2}$ × 2$\frac{1}{4}$ in.	16 oz	350–360	177–182

With the introduction of monoglyceride shortening, a three-stage blending method was initially adopted. However, during the past 15–20 years, as a result of further improvement in baking equipment and ingredients, this blending method has been further simplified by combining the first two stages, resulting in a two-stage blending method.

At first, just enough liquid should be added to give a heavy batter in which lumps will be smoothed out quickly. When the bowl and mixing paddle have been scraped down and the mix is smooth, the remainder of the liquid ingredients can be added to produce the finished batter. If all the liquid is added at once, the resulting mass is so thin that there is no rubbing action such as that obtained in a heavier dough, and the lumps are not smoothed out.

Mixing Temperature

Good-quality cakes can be made with the batter having temperatures between 60 and 80°F (15.6 and 26.7°C).

Scaling Weights and Baking Temperatures

Table 10.1 contains some typical scaling weights and baking temperatures for cakes of various sizes. Using these as a basis, the baker can determine the proper scaling weight and baking temperature for any size or shape cake he or she may wish to bake.

Handling the Finished Cakes

The cakes should be dumped from the pans a few minutes after they are out of the oven but should be cooled thoroughly before icing.

BASIC CAKE FORMULAS
(QUANTITY CAKE RECIPES)

Gold Cake

Ingredients	lb	oz	Method
Cake flour	2	8	Use a paddle and mix the
High-ratio shortening	1	12	first seven ingredients
Granulated sugar	3	2	for 5 min on the low
Salt		$1\frac{1}{2}$	speed of a three-speed
Baking powder		$2\frac{1}{2}$	mixer, or on the second
Water	1		speed of a four-speed
Nonfat dry milk		$3\frac{1}{2}$	mixer. Scrape down the
			bowl and paddle at least
			once in this stage.
Whole eggs	2	4	Scale these liquids together
Water (variable)		$12\frac{1}{2}$	and add about half to
Flavor	As desired		the above. Mix until
			smooth, scrape down,
			and mix until smooth
			again. Add the balance
			of liquids and continue
			to mix for a total of 3–4
			min in this stage,
			scraping down again to
			ensure a smooth batter

Total weight: Approx. $11\frac{3}{4}$ lbs.
Total mixing time: 8–12 min.
Baking temperature: 360–375°F (182–191°C).

Fudge Cake

Ingredients	lb	oz	Method
Cake flour	2	2	Use a paddle. Mix for 5 min
Cocoa		6	on low speed if a three-
High-ratio shortening	1	12	speed machine is used or
Granulated sugar	3	2	on second speed on a
Salt		$1\frac{1}{2}$	four-speed machine.
Soda		$\frac{3}{4}$	Scrape down the bowl
Baking powder		$1\frac{1}{2}$	and paddle at least once
Water	1		in this stage.
Nonfat dry milk		$3\frac{1}{2}$	
Whole eggs	2	4	Scale off eggs, water, and
Water		$10\frac{1}{2}$	flavor together and add
Flavor	To taste		approximately half to the bowl. Mix until smooth, scrape down, and mix until smooth again. Then add the balance of the liquid ingredients and continue mixing for a total of 3–5 min in this stage, scraping down again to ensure a smooth batter.

Total weight: Approx. 11 lbs, 12 oz
Scale: 8-in. layers (11–12 oz)
Baking temperature: 360–375°F. (182–191°C)

Yellow Pound Cake

Ingredients	lb	oz	Method
Cake flour	2	8	Use a paddle. Mix for 3 min.
High-ratio shortening	1	12	Scrape down the bowl and paddle at least once in this stage.
Granulated sugar	3		Add to the blended mass in
Salt		$1\frac{1}{2}$	the bowl and mix for 6 min
Water	1	2	scraping down at least
Nonfat dry milk		$3\frac{1}{2}$	once.

| Whole eggs | 1 | 12 | Scale off eggs and flavor |
| Flavor | To taste | | together and add |

approximately half to the
bowl. Mix until smooth,
scrape down, and mix until
smooth again. Then add
the balance of the liquid
ingredients and continue
mixing for a total of 5 min
in this stage, scraping
down again to ensure a
smooth batter.

Total weight: Approx. 10 lbs, 6 oz.

Total mixing time: 14 min.

Mixing speed: Use second speed for the first two stages if a three-speed machine is used and third speed if a four-speed machine is used. For the third stage, use slow speed on a three-speed machine or second speed on a four-speed machine, always considering the slow speed as first speed.

Baking temperature and scaling weights: Bake the 1-lb cake at about 350°F (177°C). For the 3-lb size, use 330°F (166°C).

Baking time: 1-lb cake, about 60–75 min; 3-lb cake, about 2 h.

ICING/FROSTING TECHNOLOGY

Icings are used to provide:

Eye appeal: Attractively iced cakes have more visual appeal and usually sell better than uniced cakes.

Flavor: The dominant flavor of most cakes is supplied by the icing.

Food value: The caloric food value of the icing is generally slightly greater than the caloric food value of the cake. This is because most of the icings used with layer cakes will contain between 20% and 50% fat as shortening, butter, margarine, and so on.

Moistness and freshness: An uniced cake will lose some of its desirable flavor and moistness if exposed to the air, and a "stale" condition will develop. The icing prolongs freshness.

Variety: The baker is able to produce a wide variety of cakes with a few base cakes and a few icings.

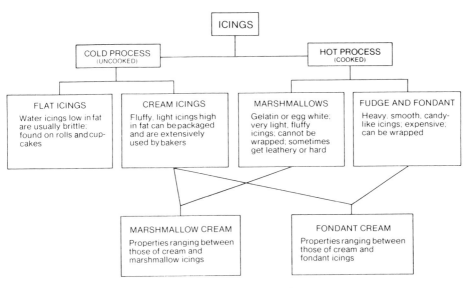

FIGURE 10.1.   Icing Classifications.

Classes of Icings

Icings are grouped in the following classes (Fig. 10.1).

Water Icing (Flat Icing)

Water icings are the simplest form of icing and consist of about 85 parts of powdered sugar to 15 parts of water with added flavor and salt. Occasionally, small amounts of other materials are added, such as shortening, corn syrup, and invert sugar. It is a dense, glossy icing used on cookies and sweet yeast goods. They are flat icings because little or no air is whipped into these icings during the mixing process.

Cream Icing

Cream icings contain 50–80% sugar, about 35% shortening, and 1–15% water. They are light and fluffy, pliable, and easily spread. The shortening enables this type of icing to take up air, which makes it light. Cream icings vary in shortening content from 20% to 55% of the weight of the icing, generally to meet a regional preference. The higher-fat cream icings are generally the preferred types in large marketing areas such as New York City, Miami, Chicago, and Los Angeles.

Fudge and Fondant Icings

Fudge and fondant icings are distinguished from other icings by the size of sugar crystals in the finished product. The crystals are small and regular and make a smooth, nongritty icing, in contrast to water icing, which has relatively large and irregular crystals and, therefore, a somewhat gritty texture.

Fudge icings are made in the same manner except that some fat is included in the boiling process. The fat retards the formation of large crystals and eliminates the need for using elaborate equipment. The boiled syrup is merely poured into a mixing bowl and agitated while cooling. When it is cooled to the proper consistency, flavor, salt, and other materials are added.

Marshmallow Icings

Marshmallow icings are made using either gelatin or egg white.

Marshmallows are essentially aerated syrups and usually contain enough water to dissolve all of the sugar crystals present so that there are no solid sugar crystals present. No fat is used in marshmallow icings, as the presence of fat destroys the whipping properties of the gelatin or egg whites. Marshmallow is light and fluffy and tends to be quite sticky.

Combinations

Combinations of many of these types of icings can be made. For example, a cream icing and a fondant can be blended to make a fondant cream. Marshmallow and cream icing can be mixed to produce an icing lighter than the cream icing.

In Fig. 10.1, this classification is shown graphically. Note that all icings are either cooked or uncooked and that the uncooked icings are divided into flat or water icings and cream icings. The cream icings can be made flat by warming or using warm ingredients. The cooked icings—marshmallow (either gelatin or egg white) and the fudges and fondants—can be used alone or in combination.

Cool the Cakes

Cakes must be cooled thoroughly before icing. Although small cakes will cool rather quickly (5–10 min), large layers or pound cakes may require as much as 30–45 min at room temperature.

In cooling, cake normally loses some moisture. If the hot cake is iced, this moisture vapor may condense between the icing and the crust, with the result that soaking is likely to occur.

Liquids

Liquid is necessary in an icing. The liquids used include water as well as water-containing materials, such as eggs, egg whites, milk, and syrups. It is the function of the liquid to dissolve some of the sugar and soften the icing so that it may be spread. A pound of water will dissolve about 2 lbs of sugar, resulting in about 3 lbs of syrup. One pound of eggs will dissolve about $1\frac{1}{2}$ lbs of sugar, resulting in $2\frac{1}{2}$ lbs of liquid.

Occasionally, the liquid functions as other than a simple binder. Egg whites in marshmallow are the material that produces the permanent foam. Molasses in icings is used for both color and flavor. Whole eggs are occasionally used in creamed icings for color and smoothness. Milk provides food flavor. Corn syrup is used principally to provide gloss to icings.

STANDARD ICING FORMULAS

Cream Icing (Low Fat)

Ingredients	lb	oz	Method
4X powdered sugar	10		Use paddle. Add half the
High-ratio shortening	4		water to other ingredients
Butter	1		and mix on low speed until
Salt		2	smooth. Add the balance of
Nonfat dry milk	1	4	the water and mix at me-
Vanilla	To taste		dium speed for about 10
Water	3	8	min to the desired light-
			ness, making sure to scrape
			down at least once.

Total weight: Approx. 20 lbs.

Chocolate Fudge Icing

Ingredients	lb	oz	Method
High-ratio shortening	1		Melt shortening and butter.
Butter		3	Add to cocoa and salt. Mix
Salt		$\frac{1}{2}$	together in mixing bowl.
Cocoa		12	
4X powdered sugar	5		Add to above and mix until smooth.
Invert sugar or honey		4	Mix together and add to the
Hot water[a] (variable)		14	above slowly to prevent lumping.

Total weight: Approx. 8 lbs, 8 oz.

[a]The optimum amount of water may vary slightly depending upon plant or shop conditions.

PIE DOUGH TECHNOLOGY

Pie doughs can be manufactured in large processing plants using specialized pie dough equipment. Medium to large upright mixers are also used in medium to large wholesale bakeries. At the other extreme, some small operations employ small upright mixers, or occasionally mix by hand.

In making pie dough, the predominant types of fats used are (1) all-purpose vegetable shortening, (2) blended meat fat and vegetable shortening, and (3) lard. The two types of shortening are most commonly packed in 50-lb cartons with polyethylene liners. Shortening used for pie crust should have the right body and consistency to produce a tender flaky pie crust. It should be a good, smooth consistency to blend readily with flour and water so as to avoid toughness and crumbliness. The ideal shortening should also have a bland, neutral flavor at room temperature and be able to resist development of off flavors at the high pie-baking temperatures employed (375–450°F; 190–232°C).

Pie dough types can be divided into (1) flaky, (2) short mealy, (3) short flaky, and (4) roll-in crusts. The ingredients used in making pie crust are flour, shortening, water, and salt. Corn sugar/dextrose may also be used to provide color.

Generally, soft winter wheat flour is best suited for making pie crust. Most bakers prefer the unbleached, long-extraction flours because they produce a more tender crust with less shrinkage. Protein level is generally 10–11%.

Shortening comprises 25–40% of the weight of the crust. Most quality bakers use 60–75 parts of shortening to 100 parts of flour. As the proportion of shortening is increased, the tenderness and flakiness of the crust are also increased. With this much shortening, any off flavors can be easily detected; therefore, the quality of the shortening is of utmost importance.

Makeup

The first three types of pie dough mentioned are made using the same ingredients, the specific type produced depending on the mixing procedure:

Ingredients	lb	oz
Pastry flour	12	8
All-purpose vegetable shortening	9	8
Salt		8
Waters (variable)*	3	
Yield: 50 shells		

*Water will vary slightly depending upon such factors as flour strength, mixing equipment, temperature, and so on.

Flaky Crust

Mix all the shortening with all the flour to obtain an irregular mixture, with little lumps of fat throughout. Add cold water and mix lightly. This dough requires slightly more water than other doughs.

Short Mealy Crust

Mix all the shortening thoroughly with half the flour. Add balance of flour and mix it in to break up the creamed mass. Add cold water and mix just enough to incorporate it.

Short Flaky Crust

Mix half the shortening with all the flour until the fat is distributed. Add balance of shortening and mix in lightly, so that it is left in little lumps throughout the dough. Add cold water and mix lightly.

Roll-in Crust

This crust may be used for making a very flaky pie crust that is suitable for the top crust of two-crust pies.

Ingredients	lb	oz	Method
Pastry flour	5		Mix first three ingredients
All-purpose vegetable	1		together well.
shortening			
Salt		$2\frac{1}{2}$	
All-purpose vegetable	2	4	Mix into above in large
shortening			lumps.
Ice water[a] (variable)	2		Mix into above lightly.
Yield: 20–21 shells			

[a]Optimum water level will depend on such factors as flour strength, mixing equipment, temperatures, and so on.

Methods of Mixing

Correct blending of the flour and the shortening is necessary if the finished product is to have the desired characteristics. If the shortening is not blended with the flour sufficiently, the crust will shrink and be tough. In such cases the flour is not coated sufficiently with the shortening to prevent the water from combining with the flour, which will overdevelop the gluten during mixing.

Overmixing after the water has been added should also be avoided. By mixing only long enough to get the water into the dough, the danger of overdeveloping the gluten is avoided, which would cause the crust to be tough.

A third important point involves the amount of water incorporated in the dough. Too much water will cause the crust to be tough; add only enough water to bring the flour and shortening mass together.

After the water has been incorporated, roll the dough out to a thickness of approximately $\frac{1}{4}$ in. and fold one-third of the dough over half of the remaining two-thirds. Then fold the remaining third on top of the first two layers. Roll out again to a thickness of $\frac{1}{4}$ in., fold to give three layers again, and roll out to a thickness of approximately $\frac{1}{2}$ in.

Then cut the dough into pieces that, when rolled out thick, will be the approximate size of the top crust of the pie. These pieces can be put on pans or waxed paper in the refrigerator and used as needed.

Baking Temperatures

Baking temperatures depend on the type and size of the pie to be baked. Pies such as pumpkin and custard are baked at about 400°F (204°C). Berry and fruit pies are baked at about 425–450°F (218–232°C). When making fruit pies, the fruit is slightly precooked in advance before adding to the pie, so the object is simply to bake the crust before the cooked filling has reached the boiling point.

Refrigerating the Pie Dough

After the pie dough is mixed, many bakers refrigerate it to increase the flakiness and to make the dough easier to roll. After the dough is refrigerated, it is cut into pieces of 8–10 oz and rolled out to the size of the pan to be used. If it is to be a fruit pie, the fruit is added at this point and the top crust is placed on top. If it is to be a custard-type pie, the unbaked filling is poured into the trimmed crust and the pie is baked. When making cream pies, the crust is baked first and the cream is added later.

The physical characteristics of shortening have an important bearing on the quality of the pie crust. A soft, oily shortening coats the flour so thoroughly that little water can be incorporated, with the result that the dough will be crumbly and difficult to roll out. On the other hand, if the shortening is too firm, it will not blend readily with the flour, and a tough crust may result.

COOKIE TECHNOLOGY

Classification

Some cookie classification terms can be confusing: Wholesale bakers may term them biscuits or hardsweets, whereas retail bakers may call them cookies. Basically, there are two general types of bakery cookies:

1. Machine-deposited such as wire-cut cookies, vanilla wafers, and so on; also, rotary type machine-cut cookies which are made in large cookie or biscuit manufacturing plants.
2. Hand-deposited by bag and tube such as drop cookies and butter cookies, and so on; hand-cut cookies such as sugar cookies, icebox cookies, and so on.

Makeup/Baking

Proper baking is a most important step in making good cookies. Unless care is taken to bake the cookies at the right temperature for the correct length of time, a cookie poor in flavor, eating qualities, and appearance may be produced even though a good formula has been used, quality ingredients have been employed, and the dough has been handled carefully. Use the information in the following paragraphs as a guide in baking cookies.

Preparation of Pans

Cookies should be baked on pans that have been:

1. Greased and floured, for cookie doughs high in moisture
2. Greased, for average rich doughs (10–15% shortening)
3. Left ungreased, for doughs rich in shortening such as butter cookies or shortbreads, which contain over 20% shortening by weight

Careful attention to the condition of the pans before the cookies are dropped on them may eliminate trouble later on with sticking and uneven baking. Cookies may stick to the pan when:

1. Pans are not clean.
2. Pans are wet or have wet spots.
3. New pans that are not conditioned properly in oven before being used. This is similar to conditioning a griddle. See Chapter 8.
4. Pans are not thoroughly greased (or greased and floured).
5. Pans are uneven on bottom (battered or bent pans).

Cookies may receive an uneven bake if the pans are badly battered or bent, as the heat from the oven will not reach the cookies on such pans at a uniform rate.

Teflonized pans are now used to a certain extent, especially for household baking.

Baking Temperature

Most machine cookies are baked at temperatures between 375 and 400°F (191 and 204°C). Cookies should be baked with a good solid heat. Flash heat should be avoided. Double panning is often essential in ovens where bottom heat is excessive and cannot be easily controlled. It is always good practice to underbake cookies slightly, as there is enough heat in the cookie pans to complete the baking process after taking them from the oven.

Shortening

Cookie varieties contain from 10% to 30% shortening, with an average of 15–20% shortening. This is based on the total dough or batter weight.

Retail bakery cookies were originally mixed with elaborate creaming methods whereby the sugar and shortening were creamed first and the rest of the ingredients were added in stages, with flour and part of the leavening last. With improved shortenings, the single stage method for mixing retail cookies was introduced. Single-stage mixing is now quite widely used in retail baking operations, but the creaming method of mixing is still employed to a great extent in wholesale cookie operations.

Shortening performance characteristics are as follows: Shortening adds caloric food value; provides tenderness, keeping qualities, grain, and texture; and adds richness to cookies.

TYPICAL COOKIE FORMULAS

Oatmeal-Raisin Cookies

Ingredients	lb	oz	Method
Vegetable shortening	1	10	Scale all the
Granulated sugar	3	8	ingredients in the
Soda		1	mixing bowl and
Salt		$1\frac{1}{4}$	mix at medium
Cinnamon		$\frac{1}{8}$	speed to a smooth
Flavor	To taste		dough (about 1–2
Oatmeal (whole)	1	4	min).
Raisins (ground)		8	
Cake flour	3	4	
Water (variable)	1		

Total weight: Approx. 11 lbs, 8 oz.
Makeup: Scale the dough into 16-oz pieces. Mold and roll by hand into
   round strips 16 in. long. Cut into 24 equal pieces, place on lightly
   greased pans, and flatten by hand or cookie stamp. This mix can also
   be used for hand-cut and machine cookies.
Scaling: Scale the cookies about 8 oz/doz.
Baking temperature: 350–360°F (177–182°C).

Butterscotch Squares

Ingredients	lb	oz	Method
Vegetable shortening	2		Scale all the
Brown sugar	5		ingredients in the
Salt		1	mixing bowl and
Pecans		10	mix at medium
Baking powder		$\frac{1}{2}$	speed to a smooth
Pastry flour	3	2	dough (about 1–2
Whole eggs	2		min).
Nonfat dry milk		1	
Water (variable)		7	
Vanilla	To taste		

Total weight: Approx. 13 lbs, 8 oz.
Makeup: Spread about 6 lbs of the above batter in a well-greased bun
   pan.
Scaling: Cut baked sheets into squares scaled 10–12 oz/doz. cookies.
Baking: Bake at 375–400°F (191–204°C). To enhance the chewy
   characteristics, keep bake on the light side.

Sugar Cookies

Ingredients	lb	oz	Method
Vegetable shortening	3		Scale the ingredients
Granulated sugar	4		into the mixing
Salt		$1\frac{1}{2}$	bowl and mix at
Mace		$\frac{1}{4}$	medium speed to a
Cake flour	5	8	smooth dough
Baking powder		3	(about 1–2 min).
Whole eggs	1		
Nonfat dry milk		2	
Water		14	

Total weight: Approx. 14 lbs, 12 oz.
Makeup: Scale the dough into 16-oz pieces. Mold and roll by hand into
round strips 16 in. long. Cut into 24 equal pieces. Place on lightly
greased pans and flatten by hand or cookie stamp. This mix can also
be used for hand-cut and machine cookies.
Scaling: Scale the cookies about 8 oz/doz.
Baking temperature: 375–400°F (191–204°C).

Chocolate Chip Cookies

Ingredients	lb	oz	Method
Vegetable shortening	2		Scale all the
Granulated sugar	3		ingredients into the
Salt		1	mixing bowl at one
Soda		$\frac{3}{4}$	time and mix at
Pastry flour	3		medium speed to a
Chopped nuts	1		smooth dough. If a
Water (variable)		8	lighter-colored
Whole eggs	1		cookie is desired,
Flavor	To taste		the chocolate
Chocolate pieces[a]	3		pieces may be
			incorporated after
			the dough is
			mixed. Mixing time
			is about 1–2 min.

Total weight: Approx. 13 lbs, 8 oz.
Makeup: Drop out the cookies by hand or with large tube bag on lightly
greased pans.
Scaling: Scale the cookies approximately 8 oz/doz.
Baking: Bake very lightly at 375–400°F (191–204°C).

[a]Chocolate pieces can be purchased through bakery supply houses, or sweet or sem-
isweet chocolate can be broken up into small pieces for use in this cookie.

SWEET DOUGH TECHNOLOGY

Food processors and bakers produce practically all of the sweet yeast goods consumed. Sweet yeast goods are not made to any extent at home. The character of the product is such that it is difficult and time-consuming to make.

Coffee cakes, sweet rolls, and Danish pastries are ideal for breakfast and for morning and afternoon coffee breaks. They are popular with children for the lunch box and afternoon snacks.

Classification of Sweet Doughs

All types of sweet dough and Danish are characterized by the fact that they are leavened predominantly by yeast action rather than baking powder or mechanical means of aeration. They are further distinguished from breads and dinner rolls by being sweeter to the taste. They provide a richer eating sensation.

Standard Sweet Doughs

The regular standard sweet dough formula is given later in this chapter. The ingredients are mixed together to form a pliable dough. The dough is allowed to ferment or "rise" for a period of time then it is made up into large units (coffee cakes) or small units (sweet rolls), proofed, and baked. Finished sweet dough is generally characterized by its soft, tender eating qualities.

Danish Pastry

Generally, a dough is used that contains less shortening and sugar than regular sweet dough. The dough is mixed in the usual manner and allowed to rest for about 15 min after which additional quantities of shortening are rolled into the dough. The dough receives additional fermentation after the rolling-in operation. It is the rolling-in operation with the subsequent folding of the dough that results in many alternating layers of fat and dough. This, in turn, results in finished baked goods that are flakier and crisper than regular sweet dough items.

Makeup of Regular Sweet Yeast Goods

Makeup Methods

There are two methods of making sweet yeast doughs; the straight dough method and the sponge dough method.

The straight dough method consists of mixing all of the ingredients to a smooth dough, fermenting for the proper length of time, and then proceeding with the makeup, proofing, and baking operations.

The sponge dough method is designed particularly for those bakers who have a large-volume production of sweet yeast goods. It has a greater fermentation tolerance in that doughs can remain on the bench longer during the makeup operation without becoming "old." The sponge dough process also works well in large wholesale bakeries, where production is carried out on large machinery. In making a dough by the sponge method, the only necessary change in the total formula is a reduction in the yeast. After fermentation, the sponge is put back into the mixer and broken up with a few revolutions on the machine before adding the other ingredients. The whole mass is then mixed to a smooth dough.

Both methods of preparing sweet doughs are used, and each baker usually has his or her own ideas about which method he or she should use. More retail bakers employ the straight dough method than the sponge dough method.

Mixing

The purposes of the mixing operation in the making of sweet yeast dough are to blend thoroughly all the ingredients to form a smooth dough and to develop sufficient structure in the dough.

The easiest way is to mix sweet doughs by means of the simplified mixing method outlined in the formula in this chapter. This simple mixing method will save time and aid in attaining more uniform production. Although the mixing time varies with flour, room temperature, mixing equipment, and other conditions in each shop, it is possible to mix sweet doughs more quickly and efficiently by this method.

Ingredient and Mixing Temperatures

The ingredients of the dough should be kept at a temperature as near to 70°F (21°C) as is practical. For fermentation to start properly, it is necessary for the dough to come from the mixer at 78–85°F (26–29°C). The heat of friction during mixing will raise the temperature this much. The temperature of the water can be adjusted to either raise or lower the dough temperature, as it may be affected by the temperature of the other ingredients. For example, during hot summer months, it may be necessary to use ice water or even crushed ice.

Fermentation

Mechanics of Fermentation

Sweet doughs contain yeast, which is a living material. During the mixing operation, yeast cells are thoroughly distributed throughout the dough and begin to feed on the sugar that is present. As the yeast feeds, carbon dioxide gas is generated, which raises the dough, making it light and porous, and makes the dough slightly acidic, which is turn makes the dough more elastic. In addition to CO_2, alcohol is produced, which may help to soften the dough and make it more elastic. Most of the alcohol evaporates during the baking cycle. (In some European bakeries the alcohol is collected.)

Factors Influencing Fermentation

Fermentation in a dough will continue until the yeast is either killed by the heat of baking or becomes inactive because the fermentable materials are used up. However, the rate of fermentation depends on a number of factors, such as the following.

Temperature: The higher the dough temperature and/or the higher the room temperature, the more rapid the rate of fermentation.

Humidity: If environmental humidity is very low, the increased evaporation of water from the dough causes it to cool and crust and retard fermentation. Otherwise humidity has little effect on fermentation.

Ingredients:

Salt: Salt retards fermentation.

Yeast: More yeast increases the fermentation rate.

Flour: Strong flours of higher protein content require a longer fermentation time than weak flours and also require more water in the dough.

Sugar: Used in quantities of up to 10% of the weight of the flour, stimulates fermentation. However, a greater sugar content retards fermentation.

Makeup of Sweet Dough

Varieties

Sweet doughs are made up into a wide range of sweet rolls and coffee cakes. The variety depends on the skill of the baker. Sweet doughs made up into larger units can be handled faster than dough made up into smaller units. This is an important factor in determining the size of the dough batch.

Time on the Bench

As each sweet dough can remain on the bench for only a certain length of time before becoming "old" or overfermented, the baker must systematize his makeup operation so as to be finished before the dough becomes "old." The best practice, as a general rule, is to make up the small pieces that require more time first and the larger pieces that require less time last. Of course, additional personnel can be used to make up the dough in order to minimize the time on the bench.

Proofing

Object

During the makeup period, much of the gas is driven out of the dough, and before the dough is baked, it is necessary that the pieces rise to their proper lightness. This is accomplished during the proofing period, when the fermentation started during the mixing of the dough continues.

Proofing Conditions

"Proofing" differs from "fermentation" in that proofing is done after makeup at a somewhat elevated temperature and humidity, whereas fermentation is carried on before makeup and at normal room temperature. Proofing is carried out at a temperature ranging between 90 and 95°F (32 and 35°C) in an enclosed cabinet. The time varies from 20 to 40 min, depending on the size of the pieces, their richness, and the actual proofing temperature.

1. Smaller pieces proof faster than larger ones.
2. Leaner doughs proof faster than richer ones.
3. Doughs proof faster at higher temperatures than at lower temperatures.

Usually no special time is given for proofing a particular sweet dough. Lean doughs are proofed until they are about double in size. Rich doughs are proofed less because of their superior "oven spring" during baking; they require only a two-thirds to three-quarters proof. The height of the pieces increases by two-thirds to three-quarters.

Humidity should be high enough to prevent crusting of the pieces, as crusting is harmful to the appearance of the baked products. A relative humidity of 75–85% is usually adequate. Some proof boxes have a humidity control.

Baking

The baking temperature will range from 375 to 390°F (191 to 199°C). (The proper range for Danish will be slightly higher.) The larger units are baked at the lower end of the range owing to the fact that at the higher end they may become too brown before becoming properly baked inside.

The baking time depends on the size of the piece being baked and may vary from 15 to 30 min, with the larger pieces naturally requiring the longer times.

No steam is needed in the oven unless there is excessive top heat, and in such cases steam can be used to advantage to prevent the pieces from browning too rapidly.

Water in Sweet Dough

A question sometimes arises as to what is the optimum amount of water to be incorporated into a sweet dough. In general, as much water should be added as a dough can carry and still have good handling properties. This produces tender, long-keeping finished products. The proper amount of water depends on the types of flour used and the ratio of bread to pastry flour. Many retail bakers do not weigh their flour, which makes it even more difficult to predict the optimum quantity of water to use in a specific formula.

Makeup of Danish

Mixing

In general, the principles outlined for sweet dough can also be applied to Danish. Mixing of dough varies, depending on shop conditions and handling procedures. Best results are obtained when the dough is not overdeveloped. This depends on the baker's previous experience. Some bakers mix until the dough pulls away from the bowl; others mix only until smooth.

Dough Temperatures

The dough should come from the mixer at temperatures ranging from 65 to 75°F (18 to 24°C). This is lower than in the case of regular sweet dough.

Generally, the colder the dough is, the better the handling properties will be. A shorter mixing time produces less bowl friction as a result of which cooler temperatures are more easily achieved.

Rollin and Folding Operations

After mixing, the dough should be allowed to rest for 15–20 min, long enough to relax and loosen up, preferably in the retarder (temperature 35–40°F; 1.6–4.4°C). Then the dough is brought out and rolled out to about 1 in. in thickness. Shortening, margarine, or butter is spread over about two-thirds of the surface of the dough. The uncovered third of the dough should be folded over the middle third. The remaining third should then be folded over to form five layers of dough and shortening.

The dough is now ready for folding. First roll out the dough again to its original size, about three times as long as it is wide. Once again, a third of the dough is folded over and the remaining third is folded on top of it; the final folded rectangle is placed in bun pans and then refrigerated. This constitutes the "first fold."

After being refrigerated for about $\frac{1}{2}$ h, the dough is removed from the refrigerator and again rolled and folded. This constitutes the "second fold."

The dough is then returned to the refrigerator for another $\frac{1}{2}$ h and once again similarly rolled and folded to complete a "third fold."

Refrigeration and Freezing

Danish and roll-in doughs lend themselves nicely to retarding or freezing operations. They may be retarded either before or after makeup into individual units.

Although the retarding process offers a very convenient and economical way of handling doughs overnight, the freezing process can extend the time between makeup and baking considerably longer. Most sweet yeast goods—Danish and roll-in dough included—are best frozen in small dough pieces or made-up units that can quickly thaw before proofing and baking.

Performance Characteristics Required in the Production of Baked Goods (8)

1. Excellent consistency, to facilitate mixing and dough handling, even after storage in warm climates.

2. Excellent working properties, so that shortening can be easily spread on Danish doughs; this makes it possible for dough to be rolled and folded with a light touch, rather than requiring a heavy motion.

3. Uniformity of product, enabling the baker to standardize his or her methods and ensure achievement of the same results day after day.

4. A large amount of water can be employed in doughs, which is an economical factor as well as a factor in improving finished products.

SWEET DOUGH AND DANISH BASIC FORMULAS

Standard Sweet Dough (for 1 gal)

Ingredients	lb	oz	Method
Sugar, granulated	4		Place all the dry
High-ratio or sweet dough shortening	4		ingredients and shortening in the mixing
Salt		4	bowl. Then at slow
Bread flour	12		speed add the eggs and
Pastry flour	6		water, in which the
Nonfat dry milk	1		yeast has been
Whole eggs	3		dissolved. Continue
Water (variable)	8	4	mixing at medium speed
Yeast[a]	2		to form a smooth, well-
Flavor	To taste		developed dough that pulls free from the side of the mixing bowl.

Total weight: Approx. 40 lbs, 9 oz.
Fermentation: Bring dough from mixer at 78–85°F (25.5–29.4°C). Allow three-fourths to full rise and then fold. You will be able to adjust the fermentation time to meet your own shop conditions.
Makeup: Let dough loosen up and take to bench. Make up into the desired pieces.
Proof: Proof at 90°F (32°C) and a relative humidity of 85% until the pieces are about double in size.
Baking temperature: 375°F (191°C).

[a]If a longer time is required on the bench in making large batches, you may need to cut down on the amount of yeast used.

Danish Pastry (for 1 gal)

Ingredients	lb	oz	Method
Sugar, granulated	3		Place all the dry ingredients and the shortening in the mixing
High-ratio or sweet dough shortening	3		bowl. Then add at slow speed the eggs and the water, in
Salt		7	which the yeast has been dissolved. Continue to mix at
Bread flour	12		medium speed for
Pastry flour	6		approximately 4–5 min.
Whole eggs	4		
Nonfat dry milk	1		
Water (variable)	8		
Yeast	2		
Flavor	To taste		

| Additional high-ratio shortening or margarine[a] | 10 | Roll in additional shortening or margarine, giving the dough three rolls, with three folds to each roll. The dough should be retarded 20–30 min between each roll and fold. |

Total weight: 49 lbs, 7 oz.

Handling: Bring dough from the mixer at 70–75°F (21–24°C). Divide into approximately 10-lb pieces, place on sheet pan (roll out until each piece fills the sheet pan), and allow to rest in cool place for 30 min.

For use with sheeter: Best handling throughout mechanical makeup is obtained by keeping the doughs on the cold side. It may also be advantageous to employ somewhat wider openings between the rollers than would be used with a harder fat.

Makeup: Make up into the desired pieces.

Proof: Proof until the pieces are about doubled in size at about 85–90°F (29.4–32.2°C) and 80–85% humidity.

Bake: Bake at 390°F (199°C). After removing from the oven, wash with butter or a hot corn syrup glaze (1 qt. of corn syrup and 1 qt. of water brought to a boil). The Danish pieces may be garnished with nutmeats, coconut, or various toppings or iced with roll icing.

---

[a]Best results are obtained when shortening or margarine temperature is 70–75°F (21–24°C).

SUMMARY

This chapter included the function of ingredients, including oils, fats, and emulsifiers that are used in the most popular types of baking products. Bakery products included in this review were cakes, icings, pies, cookies, and sweet yeast goods. Proper mixing, handling, baking procedures, and typical formulas were included as needed to properly cover this subject (9).

References

1. Institute of Shortening Edible Oils. 1988. *Food Fats and Oils*, 6th ed. Washington, D.C: Institute of Shortening and Edible Oils, p. 20.

2. Swern, D. 1979. *Bailey's Industrial Oil and Fat Products*, 3rd ed. New York: Wiley-Interscience, p. 354.

3. Smith, L. and Minor, L. 1974. *Food Service Science*. Westport, CT: AVI, p. 366.

4. Smith, L. and Minor, L. 1974. *Food Service Science*. Westport, CT: AVI, pp. 74–80.

5. Smith, L. and Minor, L. 1974. *Food Service Science*. Westport, CT: AVI, pp. 80–81.

6. Sultan, W.J. 1986. *Practical Baking*. Westport, CT: AVI, pp. 44–45.

7. Swern, D. 1979. *Bailey's Industrial Oil and Fat Products*, 3rd ed. New York: Wiley-Interscience, p. 381.

8. Hegenbart, S. 1993. *Food Product Design*, August, 110.

9. Lawson, H. 1970. *Functions and Applications of Ingredients for Baked Goods*. American Institute of Baking. Chicago, IL.

Chapter 11

Doughnut Technology

THE MECHANICS OF THE DEEP-FRYING PROCESS

The basic process for the commercial frying of doughnuts is quite similar to the frying of many other types of foods. The doughnuts, whether cake or yeast-raised, must be made up properly prior to frying. The frying kettle is loaded with shortening and is heated to frying temperature. The shortening should be carefully heated to frying temperature to avoid the danger of scorching. During the frying of doughnuts, heat is transferred from the fat to the raw doughnut batter very rapidly. The doughnuts are thoroughly cooked in about 1 to $1\frac{1}{2}$ min as they are conveyed through the length of the frying kettle.

The frying temperature is maintained by such means as direct gas-fired heating tubes running throughout the fryer or electric heating units. See Fig. 11.1 for a typical fryer for making doughnuts in bakeries and doughnut shops.

The doughnuts become golden brown in color as they become cooked and absorb shortening. In cake doughnuts, the average fat level is approximately 20–25%, whereas yeast-raised doughnuts

FIGURE 11.1. Doughnut fryer and depositor used in retail bakeries and small doughnut shops. (Courtesy of DCA Food Industries Inc.)

contain 25–30% fat. Sufficient shortening must be absorbed in order to give the doughnuts their proper eating qualities and texture, but excessive absorption will result in unappetizing, grease-soaked doughnuts. Methods of avoiding this problem will be discussed in this chapter.

The shortening must be filtered regularly in order to remove excess browned dough particles, which can become charred and cause excessive smoking and a more rapid breakdown of the frying fat.

Changes in Shortening During Deep-Frying

Changes occur in the frying fat which are characterized as:

1. Darkening of the frying shortening
2. Oxidation and polymerization
3. Hydrolysis or the development of free fatty acids

Each fat and each frying system has a typical equilibrium level at which these changes are more or less stabilized. In a doughnut frying system the operation should be sufficiently efficient that it is never necessary to throw fat away.

Rate of Color Darkening

This depends on such conditions as the type of doughnuts, the equipment, and the system. For the average medium-rich cake or yeast doughnuts, the rate of color darkening of shortening will be quite low. Doughnuts that are rich in sugar such as sucrose and corn sugar (30–35%, based on flour) will cause darkening at a more rapid rate than those containing low levels. Milk solids, and egg solids (especially whole egg solids and egg yolk solids) are browning materials, and they accelerate darkening of the frying shortening.

The types of doughnuts that have the greatest effect on darkening are the French crullers, which contain high levels of eggs. As crullers depend primarily on eggs for their lightness, those containing very high levels of eggs do not require any leavening ingredient. However, most crullers contain some ammonia to develop lightness, and this ammonia and the eggs cause the frying fat to darken at a more rapid rate than other types of doughnuts.

The more efficient the frying system, the more rapid the replacement of used fat with fresh shortening. Rapid turnover creates an equilibrium condition of relatively light-colored oil in the system.

With all other conditions being equal, the higher the frying temperature, the more rapid the rate of color darkening.

Rate of Oxidation

The important factors that affect the rate or degree of oxidation in the doughnut frying operation include frying temperature, rate of turnover, and metal contamination.

For most cake doughnuts, frying must take place at 380–385°F (193–196°C), and for yeast-raised doughnuts, at 375°F (191°C). Therefore, the frying fat must be able to take these temperatures without resulting in excessive oxidation, which tends to result in a degree of polymerization that could cause the fat to foam or result in excessive gum deposits around the frying kettle at points where hot fat, metal, and oxygen in the air come together.

As with color darkening, the more rapid the replacement of used fat with fresh, the less the amount of oxidation at equilibrium frying conditions. The kettle should be of the correct size to handle the volume of doughnut frying required, without being oversized; too large a kettle results in heating too much fat in relation to the production needs. Furthermore, production schedules should be regulated so that the frying fat is not kept at frying temperatures when doughnuts are not being fried.

Metals containing copper, such as bronze or brass, should be kept out of contact with the shortening. These metals greatly accelerate the rate of oxidation.

Rate of Hydrolysis

The rate of hydrolysis or development of free fatty acids in frying fats is due primarily to the frying temperature, the amount of moisture put through the system (from the doughnuts), and the rate of turnover.

Each doughnut frying operation has its own typical free fatty acid equilibrium level, assuming there is no fat throwaway. Free fatty acid is easy to measure, and this measurement is used by some bakers and doughnut shop operators to monitor the frying process. The typical equilibrium free fatty acid level is in itself relatively unimportant, but radical changes in the levels in a specific operation could be an indication that something is going wrong or that some drastic change has been made. This could be as simple as a change in production schedule, or a sudden increase could result from faulty frying practice, such as poor filtration, dripback from the venting system, or heating without frying.

It is interesting to note that it is necessary to "break in" fresh frying fat before doughnuts of good shape can be produced. In addition to the development of some free fatty acids, it is believed that some surface active substances are formed that aid in the production of well-shaped doughnuts.

DOUGHNUT CLASSIFICATION

Cake Doughnuts

Cake doughnuts are made from a sweetened dough that is leavened with baking powder or a combination of soda and acid material. Cake doughnuts may be considered to be made of lean cake doughs. A great variety of cake doughnuts can be made by using different shapes, flavors, filling materials, and surface finishes. Packaged powdered-sugar doughnuts are the most popular type. The most typical sizes are from 10 to 12 oz/dozen.

Yeast-Raised Doughnuts

Yeast-raised doughnuts are generally made from a sweet dough fermented with yeast to obtain its leavening action or expansion. After being fermented, the doughs are sheeted or rolled to the desired height and generally cut out by an automatic cutter or a bun divider. The pieces are made into typical doughnut shapes or rolled into a variety of shapes. Then they go through a proof box and are allowed to rise before frying. Yeast-raised doughnuts may be likened to lean yeast-raised buns or coffee cake doughs. The typical size is 16 oz/dozen.

PRINCIPLES OF GOOD DOUGHNUT PRODUCTION

Some of the most important factors in making good doughnuts are (1) good prepared mixes, (2) correct water levels, (3) proper dough handling and mixing, (4) proper makeup before frying, (5) proper frying, (6) proper care of the shortening, and (7) proper finishing.

Good Mixes

Most doughnut fryers employ prepared mixes for both cake and yeast-raised doughnuts. Therefore, the predominant factor is to select a mix from a reputable manufacturer and select one of good quality. Most of the large doughnut mix makers have a number of mixes to sell of varying degrees of quality. If a specific mix does not make satisfactory doughnuts, a baker or doughnut shop operator usually goes back to the same source for a better-quality doughnut mix. As a general rule, maximum quality is obtained from a specific mix by following the directions of the manufacturer for adding proper amounts of water to the mix, proper mixing time, and proper doughnut dough temperature. Prior to use, mixes should be stored at room temperature in a dry location. Prolonged storage, especially in hot, humid locations, can result in some loss in the leavening materials, which, in turn, can result in low-volume, poor-quality doughnuts.

Correct Water Levels

The proper quantity of water, whether the dough is made from a mix or a formula, is of utmost importance in obtaining proper dough consistency. In cake doughnuts, the proper amount of water to add to the mix is generally between 37% and 40%. In the hot summer months, the water is kept on the low side of this range in order to minimize doughnut sugaring problems.

When the dough is too slack (has too much water), the doughnuts have

1. A distorted appearance
2. Large holes
3. Excessive absorption
4. Poor expansion
5. Poor grain and coarse texture
6. Improper "break" (a marked difference in the appearance of the two sides)

When the dough is too stiff (does not have enough water), the doughnuts have:

1. A rough, broken surface on one side
2. Excessive absorption (in cracks, etc.)
3. Thick crust
4. Cracks developing during frying
5. Poor expansion
6. Improper "break"

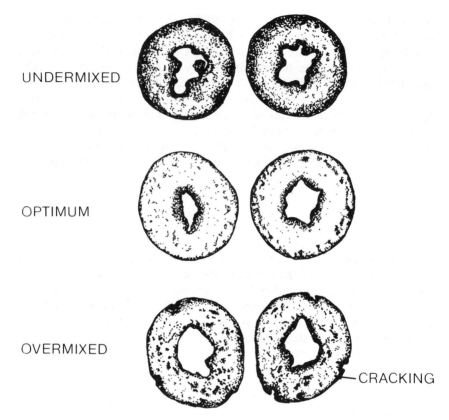

UNDERMIXED

OPTIMUM

OVERMIXED

CRACKING

FIGURE 11.2.   Effect of amount of mixing on cake doughnuts (from Ref. 2).

Proper Dough Handling and Mixing

Cake Doughnuts (Fig. 11.2)

Mixing times vary from 30 s to 8 min, depending on such factors as the following:

1. The richness of the dough: Lean mixes require less mixing time.
2. The type of flour: Weak flours or high ratios of pastry to bread flour require longer mixing.
3. The temperature of the dough: Cold doughs require more mixing than warm doughs. A dough temperature of 70–76°F (21–

24°C) is optimum. An undermixed dough will produce doughnuts with

a. Excessive absorption
b. Coarse texture
c. Irregular shape

An overmixed dough will produce doughnuts with:

a. Large holes
b. Tight grain and texture
c. Excessive absorption (due to cracking of the dough during frying)
d. A knobby, irregular surface

Proper Makeup Before Frying

Cake Doughnuts

After the correct water addition and the correct mixing, the dough is often given a 10– or 15-min rest period before cutting out. A short rest period promotes the best tolerance for the desired expansion and optimum grain and texture. The doughnuts are cut out by hand or by mechanical cutters, cutting from about 1 to 3 dozen doughnuts simultaneously. The cutters should be only a few inches above the level of shortening in the kettle in order to avoid excess splashing or damage to the doughnuts.

Yeast-Raised Doughnuts

To mix the ingredients together, first dissolve the yeast in the water. Combine all of the dry ingredients together (including the shortening), then gradually add the water while mixing at medium speed. Continue mixing to a smooth dough (about 3 min).

The optimum dough temperature in the mixer is 78–85°F (26–29°C). A fermentation time of 20–30 min is desirable before makeup. Then the dough is rolled out by hand or is fed into a series of sheeters that spread the dough and reduce the sheet thickness to accommodate the width of the proof box or sheet pans. The doughnuts can be cut by pressure and dropped directly onto the automatic proofer tray, cut out by cutters positioned over the sheet of dough, or cut out by hand. Scrap dough is fed into a takeaway conveyor, and this can be remixed into subsequent dough batches. Cutters or dies can be made to simulate figure eights, honey buns, bismarcks, or almost any size or shape. Dies can be changed in a matter of minutes.

Proofing

The proofing or secondary fermentation period for the yeast-raised doughnuts is usually carried out on wire screens or canvas sheets, regardless of how the dough has been handled up to this point. The doughnuts should be proofed a little on the "young" side. Twenty minutes is usually sufficient proofing time. Overproofed doughnuts are poor in appearance, lack full flavor, and show increased fat absorption during frying.

Proper Frying

Cake Doughnuts

For cake doughnuts, the proper frying time is about 45 s on each side, and the proper frying temperature is from 380 to 390°F (193 to 199°C). Doughnuts are flipped over at the halfway point either by hand or automatically.

Normally, cake doughnuts absorb about 3 oz of fat/dozen doughnuts. Frying at lower temperatures does not seal the surface rapidly enough, and excessive absorption results. Doughnuts are also in the kettle longer. Rich doughnuts are more susceptible to excessive absorption at low temperatures than are lean doughnuts. Temperatures above these levels prevent proper expansion and may produce soggy interiors.

Figure 11.3 shows how doughnut absorption can be affected by frying time and temperature.

The following shows the changes in composition that occur when cake doughnuts are deep fried:

	Doughnut Batter (%)	Finished Doughnut (%)
Moisture	37.5	21.0
Solids	57.5	53.5
Fat	5.0	25.5
Total	100.0	100.0

Yeast-Raised Doughnuts

Yeast-raised doughnuts should be fried at about 375°F (191°C). The basic principles of correctly frying yeast-raised doughnuts are sim-

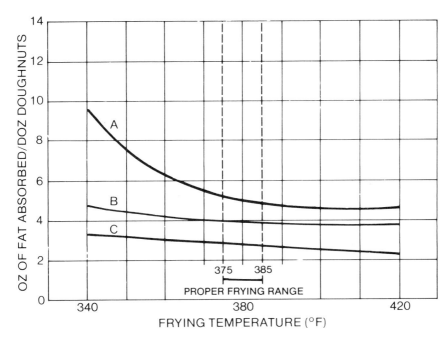

FIGURE 11.3. Absorption in relation to frying temperature (from Ref. 3).

ilar to the principles applied to cake doughnuts, except that yeast doughnuts are generally fried at a lower temperature for a somewhat longer time.

Care of the Shortening (1)

Although filtration is not the serious problem for doughnuts that it frequently is for breaded foods, a better, more efficient operation is obtained by using regular filtration. Continuous filtration may be used in large doughnut shops. The minimum filtration would involve filtering at least once per shift. Although the particles that accumulate in the fryer are quite large, they can result in excessive visual smoking of the frying fat. These particles can also promote excessive color darkening, excessive oxidation, and the development of free fatty acids. Even though these changes might not be sufficient to produce throwaways, it is preferred that the frying fat be kept in as good condition as possible to maximize both fry life and doughnut quality.

Proper Finishing

Cake doughnuts should be cooled for 20–30 min before being sugared; the temperature of the doughnuts themselves should be 85–90°F (29–32°C). If the doughnuts are too warm, the moisture from the doughnuts will soak up the sugar, resulting in a smeary surface rather than a white, snowy appearance. If the doughnuts are too cool, they will not pick up enough sugar. The most desirable sugar pickup is 18–20% of the weight of the doughnuts. In large operations, if a doughnut tumbler is used, it should be scraped at least once per shift to prevent caking or packing of sugar on the sides of the tumbler. Another important consideration is the doughnut sugar itself. Some bakers attempt to use straight powdered sugar for this operation. It is better with some shortening added and with cornstarch and milk powder incorporated into the mix. The shortening coats the sugar particles and prevents the moisture coming from the doughnut from dissolving the sugar and producing a smeary effect on the surface. The incorporation of a small amount of cornstarch provides additional moisture resistance; the cornstarch has an affinity for moisture and tends to absorb moisture coming from the doughnut before it can penetrate through the fat coating and dissolve the sugar granules. Milk powder also has a drying effect and adds some flavor to the sugar. Prepared sugars containing these materials are available from most doughnut mix suppliers.

In the case of yeast-raised doughnuts, it is customary to use a glaze consisting of sugar, water, gelatin, and flavors. Unlike cake doughnuts, yeast-raised doughnuts should be glazed hot, immediately out of the fryer. The doughnuts are conveyed through the glaze at about 100–120°F (38–49°C) (glaze temperature). The doughnuts are held until the glaze drains and sets up. Both cake doughnuts and yeast-raised doughnuts can be bagged or packed in attractive boxes.

PRINCIPAL DESIRED CHARACTERISTICS OF DOUGHNUT-FRYING SHORTENING

1. Long fry life.
2. Proper solids content at temperatures of 50–105°F (10–41°C): for excellent sugar adherence and glaze adherence for both packaged and unpackaged doughnuts.

3. Bland flavor: From 20 to 25% of the finished doughnut weight is the absorbed fat. The absorbed fat becomes an important part of the doughnut; it must be bland in flavor.

QUANTITY RECIPE SECTION

Most doughnuts produced today are made from prepared doughnut mixes. The resulting finished doughnuts vary in quality from only fair to very good, many times depending on the quality of the doughnut mix itself. Some of the more important advantages of using prepared mixes are as follows:

1. A greater degree of uniformity of production, especially with less-skilled employees.
2. A lesser chance of making an error. Errors are more likely when scaling a greater number of ingredients.
3. A certain amount of time savings in production.

FIGURE 11.4.   Intermediate capacity doughnut fryer. Production rate approx. 400 doz./h. (Courtesy of DCA Food Industries Inc.)

Some of the major disadvantages of using doughnut mixes are as follows:

1. The finished doughnut quality is similar to the doughnut quality of others who use the same prepared mix or mixes of similar quality.
2. The ability to innovate and be creative is limited to such things as doughnut sizes and shapes and innovative recipes for fillings and toppings.

Figure 11.4 illustrates an intermediate size doughnut fryer that produces about 400 doz./h. This equipment is used in large retail bakers, multi-unit bakeries, and some doughnut shops.

For very creative doughnut producers, the following section of quantity recipes for cake doughnuts and yeast-raised doughnuts, along with fillings and toppings, should prove to be helpful.

Golden Cake Doughnuts

Ingredients	lb	oz	Method
Powdered sugar	3		Mix together at
Vegetable shortening		8	medium speed
Vegetable shortening		8	(second speed of a
Salt		2	three-speed
Nonfat dry milk		10	machine, third
Soda		$\frac{1}{2}$	speed of a four-
Pastry flour	4		speed machine) for
Bread flour	4		5–8 min.
Baking powder		6	
Egg yolk	2	8	
Water (variable)	4	4	
Flavor	To taste		

Total weight of mix: Approx. 19 lbs.
Frying: Fry at 375–390°F (191–199°C).
Production pointers: This formula is set up for machine-made doughnuts.

Doughnut Sugaring Formula
(Suitable for Any Cake or Baking Powder Doughnut)

Ingredients	lb	oz	Method
Powdered sugar	8		Cream together for 10–15 min.
Shortening[a]		8	
Nonfat dry milk		8	Add and mix for several
Cornstarch	1		minutes. Run finished mix through a very fine sieve.

Total weight of mix: Approx. 10 lbs.

[a]Note: The amount of shortening in the doughnut sugaring mix can be varied; by increasing the amount in the above doughnut sugar, a heavier type of doughnut sugar will be produced.

Icing Formula (for Cake Doughnuts)

Ingredients	lb	oz	Method
Water	1	8	Bring to a boil.
Glucose		8	
Gelatin[a]		1/2	Dissolve gelatin in the water
Water		8	and add to the above syrup.
Powdered sugar	12		Add the syrup slowly to the sugar and beat until smooth. Bring to temperature of 90–100°F (32–38°C) for use.

Total weight of mix: Approx. $14\frac{1}{2}$ lbs.

[a]The amount of gelatin may be varied, depending on its strength.

Yeast-raised Doughnuts

Ingredients	lb	oz	Method
Granulated sugar		14	Place all the dry ingredients,
High-ratio shortening	1	8	including shortening, in the
Salt		2	mixing bowl, then at medium
Nutmeg		$\frac{1}{4}$	speed add the eggs and water
Mace		$\frac{1}{4}$	in which the yeast has been
Bread flour	4		dissolved. Continue mixing
Pastry flour	3	8	until a smooth, well-
Baking powder		$2\frac{1}{2}$	developed dough is formed;
Whole eggs	1		usually 8–10 min.
Nonfat dry milk		8	
Water	4		
Yeast (compressed)		10	

To make praline doughnuts, add 1 lb 8 oz chopped pecans to the above mix. Use praline icing.

Total weight of mix: Approx. $16\frac{1}{2}$ lbs.

Makeup: Bring the dough from the mixer at 78–85°F (26–29°C), let it rest for 20 min, and take it to the bench. Roll out to $\frac{1}{2}$-in. thickness. Cut out doughnuts and proof about 20 min.

Proofing: The doughnuts should be proofed a little on the "young" side; 20 min is usually sufficient proofing time. Overproofed doughnuts will be poor in appearance, will lack full flavor, and will show increased fat absorption during frying.

Frying temperature: Fry the doughnuts scaled 16 oz/dozen at 360–375°F (182–191°C).

Finishing: Properly glazed yeast-raised doughnuts can be produced by bringing the doughnuts from the frying kettle and glazing them while they are still hot. A convenient way to handle the glazing operation is to dip the doughnuts on the wire screen in the glaze and then allow them to remain on the wire screen until the glaze sets up.

Whole Wheat Doughnuts

Ingredients	lb	oz	Method
Brown sugar		10	Cream up together.
Salt		2	
Mashed potatoes (cooked)	2		
High-ratio shortening	1		
Whole eggs		10	Add and stir in.
Liquid milk	4	8	Dissolve the yeast in
Yeast		12	the milk and flavor.
Flavor	To taste		Add and mix in.
Whole wheat flour	2		Blend together, add,
Bread flour	2	8	and mix to a
Pastry flour	3		smooth dough.

Bring the dough from mixer at 80°F (27°C). Let the dough rise for $\frac{1}{2}$ h, take it to the bench, make up and proof. Fry at 360–370°F (182–188°C).

Total weight: Approx. $17\frac{1}{4}$ lbs.

Honey-glaze Doughnut (Suitable for Any Yeast-Raised Doughnut)

Ingredients	lb	oz	Method
4X powdered sugar	10		Mix to a paste.
Honey		6	
Hot water	1		
Gelatin[a]		2	Dissolve, add, and
Hot water (variable)	1		mix smooth. Bring to a temperature of 100–120°F (38–49°C) for use.

Total weight of mix: Approx. $12\frac{1}{2}$ lbs.

Note: Glaze the doughnuts by dipping them into the glaze made from the formula given above as they come from the frying kettle. Take the doughnuts from the glaze and allow them to drain on wire screens until the glaze sets up. Doughnuts can be covered with chopped nuts, coconut, or granulated sugar if desired.
[a]The amount of gelatin may be varied, depending on its strength.

SUMMARY

In this chapter we described the technology of producing both yeast-raised and cake doughnuts. Both the similarities and differences between the doughnut frying process and other fried foods were pointed out. Ingredients used, proper procedures, shortening handling, and equipment used in large and small doughnut frying operations were included in this chapter.

References

1. Swern, D. 1979. *Bailey's Industrial Oil and Fat Products*, 3rd ed. New York: Wiley-Interscience, p. 384.
2. Lawson, H. 1985. *Standards for Fats & Oils*. Westport, CT: AVI, p. 176.
3. Lawson, H. 1985. *Standards for Fats & Oils*. Westport, CT: AVI, p. 178.

Chapter 12

Other Large Commercial Uses

This chapter will cover some additional important uses for food fats and oils. It could be titled "Food Processor Uses," because it is basically an overview of the technology and utilization of fats and oils in relatively large quantities. Obviously it cannot be an in-depth study of each specific industry but should be considered a guide for use especially by food technologists in allied industries, suppliers to the food processing industries, their customers, and interested students.

MARGARINE (1)

Margarines are food spreads prepared by blending fats and/or oils with other ingredients such as water and/or milk products, salt, lecithin, emulsifier, flavoring and coloring materials, and vitamins A and D. By federal regulation, margarine must contain at least 80% fat. The emulsifier aids in maintaining a very fine dispersion of the aqueous phase in the fat after processing, as well as reducing spattering during use. The lecithin greatly lessens the possibility of sticking when the margarine is used for griddling and panfrying. The margarine industry also produces reduced calorie (or diet, imitation) margarines containing 40–52% fat and spreads containing 40–75% fat.

Margarines are molded and packaged in $\frac{1}{4}$- and 1-lb stick forms.

Table 12.1. Fatty Acid Composition of Fats and Oils in Typical Grade Margarines and of Butterfat

Product	% of Total Fatty Acids		
	Monounsaturated	Polyunsaturated	Saturated
Stick margarine			
All vegetable	45–66	14–35	18–21
Animal and vegetable	46–52	9–19	29–40
Tub magarine:			
All vegetable	33–52	29–48	17–19
Spreads[a] (margarine substitute):			
All vegetable	32–54	27–50	17–20
Butterfat[b]	28–31	1–3	63–70

[a]Spreads are margarinelike products containing less than 80% fat. Typically, such products contain 40–75% fat.

[b]Butterfat contains about 0.2–0.4% arachidonic acid. The data for saturated fatty acids include the contributions of C_4, C_6, C_8, and C_{10} saturated fatty acids, which represent about 10% of total fatty acids.

Source: Courtesy of Institute of Shortening and Edible Oils, Inc.

Many manufacturers also produce soft margarines that are high in polyunsaturated fatty acids that are not hydrogenated, and which are sold in $\frac{1}{2}$- and 1-lb tubs. To a lesser extent, these tub margarine formulations are also sold as whipped and diet products. Spreads are also available in both stick and tub form. Some manufacturers produce margarines in fluid form to provide additional convenience. The fats used in margarine may be from either vegetable or animal origin, although vegetable oils are used more widely. In the early days of margarine manufacturing, beef fat was the predominant fat ingredient, but vegetable oil types are more popular today.

The fat in margarine may be prepared from a single hydrogenated fat, from two or more hydrogenated fats, or from a blend of hydrogenated fat(s) and unhydrogenated oil(s). This offers the manufacturer a wide range of compositional flexibility. Vegetable oil margarine producers have in recent years marketed products with substantially higher levels of polyunsaturated fatty acids. Margarine fats and oils usually contain about 14–48% polyunsaturated fatty acids. Table 12.1 gives the range in fatty acid composition for the various types of margarine products available in today's retail market (2).

Every margarine manufacturer produces a wide variety of brands to meet different needs and wants of their customers. Specifications for the fat/oil portion of each product are prepared by the margarine maker. These specifications are given to their refiner/supplier. The supplier will refine, bleach, hydrogenate (as required), and deodorize the oils as specified. Each product will be completely formulated including the source oils specified and will be shipped to the margarine manufacturer in railcars. They are usually kept under heat to above their complete melting point [120°F (49°C) or higher] and under nitrogen blanketing. At the margarine plant, this formulated fat is mixed with the other ingredients (water, milk products, salt, etc.) in the proper proportions for the specific finished product.

The freezing or solidification process is somewhat different than that employed in making shortening. The fact that margarine contains only 80% fat, with the remaining 20% being predominantly water, does not present any undue complications in this solidification process. Most margarines are solidified in continuous freezing (or Votator) processing equipment. In the first stage, the product goes through typical shortening-type freezers, generally referred to as "A" units. A typical installation would have two or three of these "A" units lined up in series. The margarine mixture enters the "A" unit at about 100°F (38°C) and is chilled to about 50°F (10°C) with a residence time of about 16 s. The "A" units have similar mechanical agitation as the shortening freezers. Coming out of the "A" units, the product is quite fluid because it is in a supercooled condition. The product then goes through a "B" unit which does not have mechanical agitation. The reasons for limiting the amount of work or agitation in the "B" unit include:

1. The product must not be too soft for handling in the automatic print-forming and wrapping equipment.

2. To prevent the water phase from being dispersed in too fine a state of subdivision.

3. To induce the growth of somewhat larger crystals as compared with shortening.

Larger crystals in margarine are desirable in order to obtain a faster meltdown, or a less "waxy" feeling in the mouth. Also, too fine a crystal structure results in brittleness or hardness and a lack of spreadability at refrigerator temperatures.

The supercooled mass from the "A" units solidifies as it is slowly forced through the "B" unit. The use of two "B" units in parallel is a typical installation. When one "B" has been filled, the flow of product is switched to the second unit. The product in one section remains static until the second section is filled. The solidified margarine is then extruded and moved on to the packaging section (3).

COCOA BUTTERS AND HARD BUTTERS (4)

Cocoa butter is the most important of the vegetable butters in the United States. The major uses are in the manufacture of chocolate confections and in coatings for chocolate and candies. Also, it is used to a certain extent in cosmetics and suntan preparations. Most of the world's cocoa comes from Africa, and it is obtained from the tropical plant Theobroma Cacao.

Cocoa butter has the unique property of melting very sharply with softening and melting over an even more restricted range than coconut oil or palm kernel oil. The short plastic range of cocoa butter is not due to the presence of low-molecular-weight fatty acids but from an unusual triglyceride composition. They consist mainly of relatively simple mixtures of triglycerides, either of a single triglyceride or of two or more triglycerides of very nearly identical melting points. This composition not only causes the fat to melt sharply but also confers distinctive habits of crystallization. Conditions under which cocoa butter has been solidified and stored have a great effect on its consistency and melting point.

The typical fatty acid composition of cocoa butter is palmitic 24–27%, stearic 33–35%, oleic 34–37%, and linoleic 2–4%. The complete melting depends on the crystal modification. In its highest melting form it will be completely melted at 95–97°F (35–36°C), which is just below normal body temperature.

As the characteristic flavor of cocoa butter is desirable in confections and chocolates, it is generally not subjected to refining, bleaching, and deodorization, unless its quality has suffered from age or from damage to the cocoa beans. Compared with most fats and oils, cocoa butter is extremely resistant to deterioration through oxidation or the development of rancidity.

Cocoa butter is also referred to as a "hard butter."

Due to the relatively high price of cocoa butter, it is subject to adulteration with other fats. In addition, with its high price and difficulty in obtaining a continuous supply at times, there have been extensive efforts to develop replacements or extenders. Although to date no completely equivalent confectionery fats have evolved, they are getting better and better. Replacements have been manufactured to a great extent through fractionation, primarily using coconut oil as a base oil.

Other hard butters and cocoa butter replacements are fats that exhibit high solids content at room temperature and below (50°-70°F; 10°-21°C) but that melt rapidly at body temperature (98.6°F; 37.0°C). Cocoa butter, the fat present in chocolate, is a hard butter. Some vegetable fats, such as cocoa butter, have hard butter properties in their natural state; but most vegetable fats and oils, including soybean, cottonseed, coconut, palm, palm kernel, and shea, must be modified by processing to achieve hard butter characteristics. Shea butter is obtained from the kernels of an African plant named *Butyrospermum parkii*. Typical processing to derive hard butter properties includes one or more of the following steps: blending of oils, random and directed interesterification, fractional crystallization, and hydrogenation.

Hard butters are used to formulate a variety of convenience foods such as nondairy products and confectioners' coatings.

Over the years, there have been many research projects designed to develop other vegetable fat products to replace cocoa butter. In the past they have been based primarily on economic or price objectives. In recent years, the goals have been both economic and nutritional. Recently, Procter & Gamble announced the development of a reduced-calorie fat that has the performance characteristics of cocoa butter. It has fewer calories because one of its three fatty acids is only partially absorbed by the body. It has been submitted to the Food and Drug Administration for review and approval.

DAIRY REPLACEMENTS

In addition to the extensive use of margarine, vegetable fats and oils are used to replace butterfat in almost all of the basic dairy products

DAIRY PRODUCT	REPLACEMENT
Butter	Margarine
Whipped cream	Whipped topping
Ice cream	Mellorines/frozen desserts
Coffee cream	Coffee whiteners
Sour cream	Imitation sour cream
Cheese	Imitation cheese/cheese spreads
Ice milk	Frozen desserts
Milk	Milk replacements

Standard dairy products are produced under Standards of Identity as promulgated by the Food and Drug Administration (5). Products such as butter, ice cream, and ice milk require a minimum amount of butterfat, 80%, 10%, and 7%, respectively. The FDA is considering changing the standards for both ice cream and ice milk. The International Ice Cream Association has suggested permitting use of the name "reduced-fat ice cream" for what is currently known as "ice milk" and be allowed to have a maximum butterfat content of 5% instead of 7%. Another possible resolution would be to completely eliminate the requirement of 10% butterfat in ice cream and have the manufacturers provide the fat content on their labels.

Reduced-butterfat cheese products are quite well established in the marketplace, although their quality in relation to the comparable regular butterfat products is questionable. These cheese products range from a 33% reduction in butterfat on down to even a zero fat level in some products.

Low-fat and nonfat milk products are also well established, and a good segment of consumers appear to be satisfied as they develop a taste for the lower-fat products.

When formulating the products with vegetable fats and oils to replace butterfat, some considerations must apply. In addition to nutritional issues, major product characteristics to consider include texture, lubricity, flavor, aeration properties, appearance, and stability (or shelf life).

Texture and structure are related to mouth feel. Coconut oil and palm kernel oils have been used in applications requiring a sharp melting curve, and this is especially desirable in products such as coffee whiteners. When further replacing these oils with domestic oils, it is generally necessary to give up some of these mouth feel characteristics.

Liquid oil lessens the abrasive effect of other ingredients during mixing. Oils that are liquid between refrigeration temperatures and room temperatures are desirable as milk replacements.

Liquid oil inhibits incorporation of air and the development of a stable structure in shortenings and other food products (especially without added emulsifiers). Aeration is promoted by the fat crystals in properly processed plastic shortenings. For whipped toppings, frozen desserts, and other imitation ice creams, a shortening made from partially hydrogenated soybean oil (beta-crystal-tending) with some palm or cottonseed (beta-prime-tending) works well in this regard.

LARGE COMMERCIAL BAKING

Very large-scale wholesale or commercial baking operations are significant users of various types of fats and oils. Some of the operations included in this list are standard white bread, variety and specialty breads, rolls, pies, cake and pound cake, fillings and coatings, crackers, pretzels, cookies and biscuits, sweet yeast coffee cakes and rolls, Danish pastry, puff pastry, ice cream cones, and prepared mixes.

Some companies are single-plant operations, but more and more they may be characterized as large multiplant corporate entities. The basic principles of baking and the function of ingredients follow the rules covered earlier in Chapter 10. The main difference is in the size of the operation and the bulk handling of major ingredients. Specially designed rail cars are employed for receipt of flour. Trucks of 20,000–60,000-lbs capacity and rail cars of up to 150,000-lbs capacity are used for fats and oils. Liquid oils and fluid shortenings are transferred under atmospheric temperatures without the application of heat. Liquid salad and cooking oils do not require added heat for pumpability, and the added heat would tend to reduce the products' oxidative stability. In the case of fluid shortenings, this heat and subsequent cooling to room temperature would result in the product setting up in a gellike consistency, thus nullifying one of the most important advantages for fluid shortening. However, shortening and margarine-type products must be shipped under circulating heat to keep these products above their melting point (120–150°F; 49–66°C). If a plastic shortening is required, it is generally shipped completely formulated as per specification, and then it is processed through freezers and plasticized before use in the plant. Rail shipments of fats and oils, especially those under heat, are transported under nitrogen blanketing in order to minimize oxidative changes.

Large commercial bakery plants are also characterized by the employment of more and more continuous automatic equipment under computer control.

Figure 12.1 illustrates a typical flow diagram for a continuous cake production line. Figures 12.2 and 12.3 illustrate typical batch bread and continuous bread production.

LARGE COMMERCIAL FRYING

This group includes large commercial doughnut plants; potato chip, corn chip, and other fried snacks; egg roll and Chinese commercially

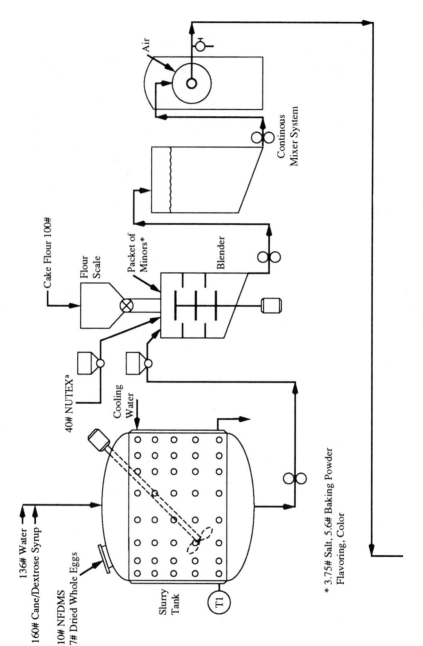

FIGURE 12.1.   Continuous Cake Production

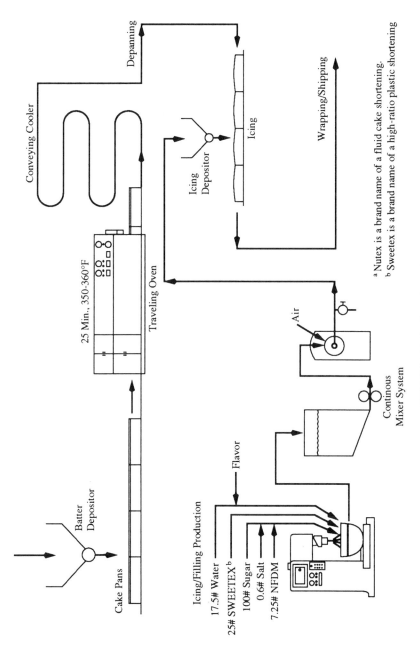

FIGURE 12.1. Continued

[a] Nutex is a brand name of a fluid cake shortening.
[b] Sweetex is a brand name of a high-ratio plastic shortening.

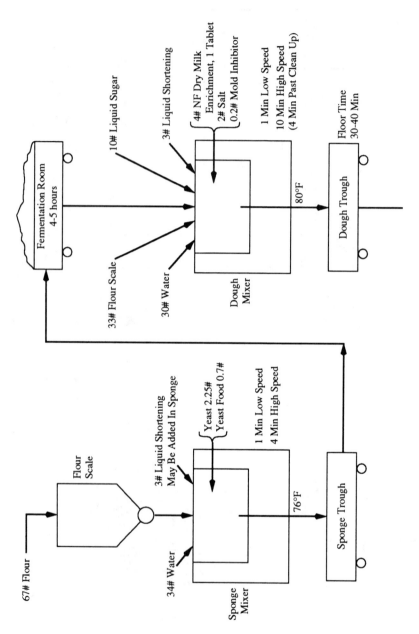

FIGURE 12.2.   Batch Bread Production

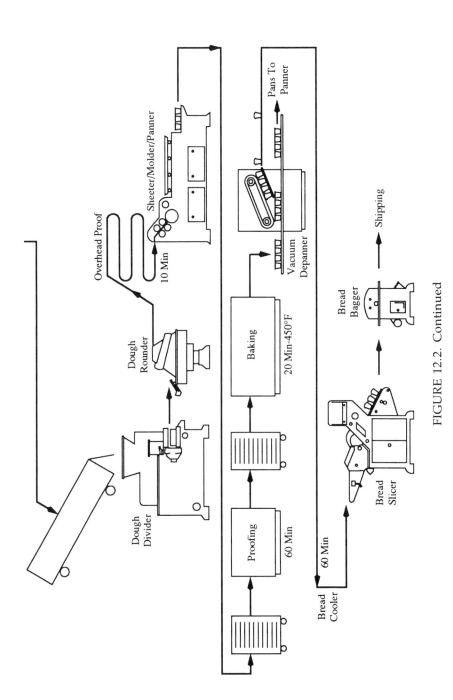

FIGURE 12.2. Continued

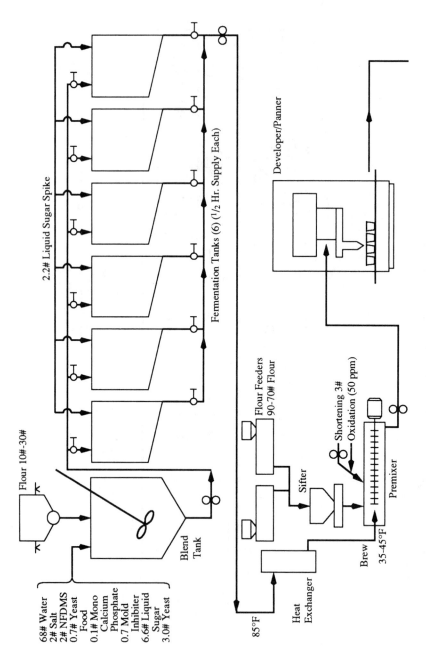

FIGURE 12.3.   Flow Chart for Continuous Bread Production

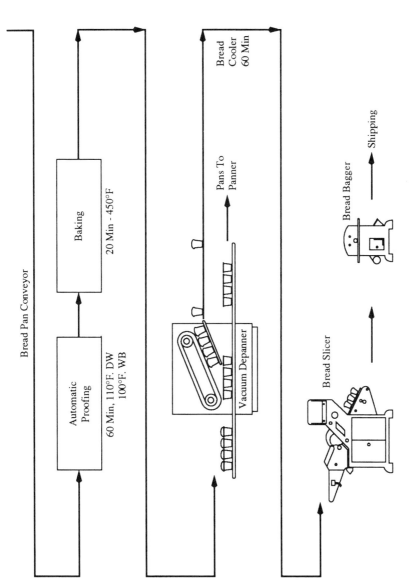

FIGURE 12.3. Continued

FIGURE 12.4.   High-capacity doughnut fryer. Production rate approx. 1800 doz./h. (Courtesy of DCA Food Industries Inc.)

fried foods; and prefried and frozen poultry products, fish and other seafoods, and meats.

Again the basic principles of deep-frying given in Chapters 7 and 11 apply in this type of frying. It is merely a matter of applying good basic frying principles, using common sense, and adapting this to large-scale production.

Equipment used for the frying operation is considerably larger than foodservice equipment, holding from a few hundred pounds of frying fat or oil up to several thousand pounds. In general, the equipment is manufactured by different equipment manufacturers than found in foodservice. Much of the equipment is designed and manufactured to fit the specific needs of the customer. Many large commercial food processors design and manufacture their own equipment.

In most operations, frying is continuous, computer operated, and sufficiently efficient that frying fat is never discarded. Automatic and continuous or semicontinuous filtering contributes to this efficiency.

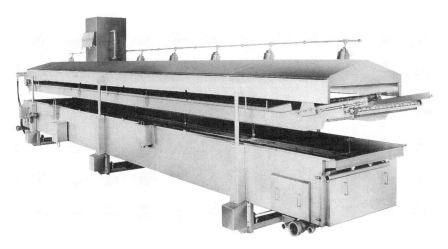

FIGURE 12.5.   Direct fired commercial food fryer. (Courtesy of Heat and Control, Inc.)

The frying fat can utilize gas, electric, or oil to heat the system, taking advantage of fuel economies in certain areas. The oil or fat may be heated by direct contact with electric immersion coils, by gas-fired hot air in tubes circulating through the frying fat, or by indirect heat transfer. In the latter, the frying oil can be heated outside the frying machine and circulated back through the frying chamber, or a heat transfer fluid can be circulated from the heater through the frying fat and back to the heating equipment.

Conveyors or conveyor belts may be used to move the food into the frying fat, through the kettle, and on to an exit conveyor.

For foods that are to be frozen, they must be cooled rapidly for warehousing, distribution, and marketing. Usually the food is quick frozen by liquid nitrogen or blast freezing.

Figures 12.5 and 12.6 illustrate the safety and performance features of a very good commercial food fryer. It is direct fired by totally sealed burners and firing tubes that produce uniform heat throughout the fryer. Other quality features include automatic frying oil level sensors, continuous removal and filtration of fines (crumbs), and mechanical screw-jack hoists to lift the cover. Safety features include (1) a frying oil level control system with an automatic low-level shut-off, (2) a stack overtemperature sensor with automatic burner shut-off, (3) a wall-mount control panel that includes all of

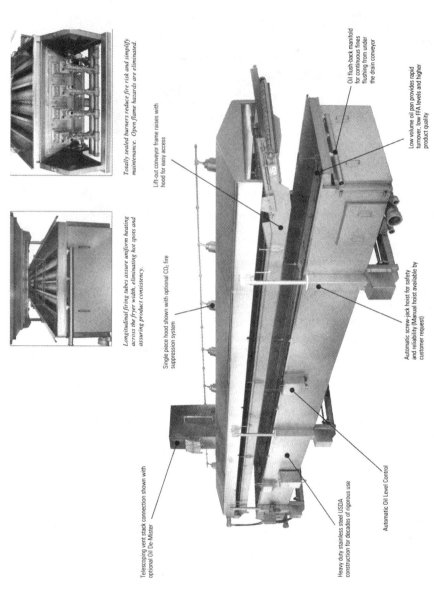

Totally sealed burners reduce fire risk and simplify maintenance. Open flame hazards are eliminated.

Longitudinal firing tubes assure uniform heating across the fryer width, eliminating hot spots and assuring product consistency.

Lift-out conveyor frame raises with hood for easy access

Oil flush-back manifold for continuous fines flushing from under the drain conveyor

Low volume oil pan provides rapid turnover, low FFA levels and higher product quality

Single piece hood shown with optional CO₂ fire suppression system

Automatic screw-jack hoist for safety and reliability (Manual hoist available by customer request)

Telescoping vent stack connection shown with optional Oil De-Mister

Heavy duty stainless steel USDA construction for decades of rigorous use

Automatic Oil Level Control

FIGURE 12.6.   Direct fired commercial fryer illustrating safety and efficiency features. (Courtesy of Heat and Control, Inc.)

FIGURE 12.7.   Snack food processor; east side view.

the operating controls and digital temperature indicators, and (4) a CO_2 fire suppression system.

Another type of commercial food frying is illustrated in Figs. 12.7–12.9. This machine was designed to process prefabricated potato chips.

SUMMARY

In this chapter we have covered some of the larger commercial food operations in which oils and fats play a major role.

PARTIAL VIEW – WEST SIDE

FIGURE 12.8.   Snack food processor; west side view.

MAIN DRIVE UNIT, LOCATED ON EAST SIDE

FIGURE 12.9.   Snack food processor; main drive unit.

References

1. Institute of Shortening Edible Oils. 1988. *Food Fats and Oils*, 6th ed. Institute of Shortening and Edible Oils, Washington, DC: pp. 21–22.
2. Institute of Shortening Edible Oils. 1988. *Food Fats and Oils*, 6th ed. Institute of Shortening and Edible Oils, Washington, DC: p. 22.
3. Applewhite, T. 1985. *Bailey's Industrial Oil and Fat Products, Vol. 3*, New York: Wiley-Interscience, pp. 77–86.
4. Swern, D. 1982. *Bailey's Industrial Oil and Fat Products, Vol. 1*, 4th ed. New York: Wiley-Interscience, pp. 322–332.
5. USDA. 1971. *Federal and State Standards for the Composition of Milk Products and Certain Non-Milkfat Products*. Washington, DC: U.S. Government Printing Office.

Chapter 13

Nutritional Aspects of Oils and Fats

Fred J. Baur, Ph.D.

INTRODUCTION

In this chapter the words "fat" or "fats" are used to designate oils and fats as these terms are used elsewhere in this book.

Fats are a principal and essential constituent of the human diet along with carbohydrates and proteins. Fats are a major source of energy supplying about 9 kcal/g. Proteins and carbohydrates each supply about 4 kcal/g.

In calorie-deficient situations, fats together with carbohydrates spare protein and improve growth rates. Some fatty foods are sources of fat-soluble vitamins (A, D, E, and K) and the ingestion of fat improves the absorption of these vitamins regardless of their source. Fats are vital to a palatable and well-rounded diet and provide the essential fatty acids linoleic and linolenic. Fats are also essential components of all cell membranes.

In the second half of this century, emphases on nutrient-related diseases increased as the major causes of mortality shifted from infectious to chronic diseases. Attention turned to investigating the role of diet in the maintenance of health and reduction of risk in such chronic diseases as heart disease, cancer, hypertension, obesity, diabetes, and others.

Just as necessary drugs have contraindications relative to their uses, contraindications frequently exist in the selection of appropriate diets. For example, enhanced calcium intake for possible protection against osteoporosis via ingestion of dairy products is contraindicated by their being rich sources of saturated fatty acids and cholesterol, which when increased have been related to the risk of coronary heart disease (CHD). Two recent press releases relative to alcohol consumption provide an example relative to caloric intake. One headline said "Study: Drinking may lower risk for heart attack" (1) and the other, "Two drinks a day increase risk of breast cancer" (2).

Absolute proof is difficult to obtain in any branch of science. Instead, accumulated evidence frequently reaches the point that proof is accepted on operational bases. This is where the role of fats in the human diet is and will likely remain; that is, proof positive of the effects of fats in many aspects will remain lacking.

Advances in nutritional research and increasing public preoccupation with personal health and fitness have led to much popular and scientific debate on the relative merits of various dietary fats. The questions are becoming more, not less, complex. Much of the available information may seem contradictory and confusing to scientists as well as consumers. Some food suppliers have capitalized on the confusion, labeling products such as bananas and corn oil "cholesterol free" (3).

In the laboratory, scientists are sorting out the data: Some dietary fats when substituted for others lower blood cholesterol levels; other dietary fats raise blood cholesterol levels. Dietary polyunsaturated fats may enhance the growth of colon and breast cancers; yet one conjugated polyunsaturated fatty acid has been shown to be a potent inhibitor of cancer formation. Omega-3 fatty acids found in fish oil may lessen the severity of some autoimmune diseases. Research on dietary fat consumption and obesity is beginning to show positive results. Other scientists are concerned with fat requirements of infants. Nutritionists and food companies are studying the effect of fat substitutes and alternatives on overall fat consumption and health. Meanwhile, these issues are clouded by the likelihood that different people are genetically programmed to respond differently to dietary fats (3).

LEADING CAUSES OF DEATH IN THE UNITED STATES

Of the 10 leading causes of death in the United States, 6 have diet as a factor. These are heart disease, cancer, stroke, diabetes, liver disease, and atherosclerosis.

Many believe that cancer is the number one killer in the United States. That is incorrect (4). Cardiovascular disease (CVD) is by far in the lead. In 1990, heart and blood vessel diseases killed about 944,000 Americans, almost as many as cancer, accidents, pneumonia, and all other causes of death combined. In fact, coronary heart disease (CHD) alone caused 489,300 deaths in 1990, almost as many as cancer from all causes, about 506,000 (5). Stroke, another cardiovascular disease, was in third place at 145,340. Whereas the death rate from CVD has declined 32.6% in the 1980–1990 period, cancer mortality has had a steady rise with 526,000 deaths predicted for 1993 (5), with lung cancer clearly in the lead as it has been since about 1965.

Fats (lipids) are implicated in CVD and that implication is the subject of much of this chapter. The evidence for relation of fats to cancer is less convincing, but data do exist for concerns relative to the colon/rectum, prostate, and the pancreas. Prostate cancer is the second leading cause of death due to cancer in males, second only to lung. Approximately, 1 in every 10 men will develop prostate cancer by age 85 and this is likely to increase as males live longer (5). Skin cancer (both sexes) leads with over 700,000 new cases per year; most are the highly curable basal cell and squamous cell cancers. New melanoma cases will run about 32,000. Since 1973, the rate of increase has been about 4% per year. Further increase is anticipated because of the loss of the ozone layer.

CALORIC INTAKE/OBESITY

Over 25% of Americans are classified as overweight or obese. Overweight individuals tend to have higher total cholesterol, higher LDL lipoproteins (considered the bad cholesterol) and triglycerides, and lower HDL lipoproteins (considered the good cholesterol). The greatest differences have been in men under 45 years of age. A study of the data from the United States National Health and Examination Survey (1976 through 1980) yielded the conclusion that obesity is epidemic in the United States (6). An "absolute majority" of U.S. white males are greater than 12% overweight. Whereas public policy on coronary heart disease and other chronic diseases has advised action on obesity, the main recommendations have dealt with dietary lipid composition and cholesterol rather than total fat. A companion article (7) speaks to the influence of body weight on serum lipids. Overweight individuals tend to have higher triglycerides, lower HDL (good cholesterol), and higher non-HDL (principally LDL— bad cholesterol) and total cholesterol.

Obviously, a key need for ingesting food is to meet energy and growth requirements. Interestingly, the scientists working on this important, complex field in recent years do not associate excess fat ingestion directly to obesity. A Harris poll in 1993 showed that 66% of Americans have a weight problem (8). It is known that calories unused for energy and growth purposes are stored as fats and that fats are second to carbohydrates as a major source of calories in present U.S. diets. It seems, therefore, that there is a direct association between excess fat intake and obesity and will remain so even when the intake level of fats is dropped as is presently recommended by the major health professionals and professional organizations. The attitude that has changed over the past decades is that weight gain or loss is much more complex than a simple caloric intake versus needs or that obese individuals are gluttons. Whereas overeating can lead to obesity, the weight gain cannot be accounted for solely by the increased energy intake. Other key factors can be reduced physical activity, altered metabolic efficiency, and genetic predisposition.

What evidently is known is that:

1. Overweight individuals eat more than they think, perhaps twice as many calories; a recent study indicated that the failure of some obese subjects to lose weight was not only reporting too few calories but also overestimating physical activities (9).

2. Eating a high-fat diet is conducive to obesity. A person will get fatter on a high-fat, low-carbohydrate diet than on a low-fat, high-carbohydrate diet, even if both diets supply the same number of calories.

3. People discontinuing weight-reducing diets have a greater preference for high-fat diets (satiety factor?) which may explain the prompt swings in weight gain and loss, the so-called yo-yo effect.

4. Neither frequent fluctuations in body weight nor extreme restrictions of food intake is desirable (10).

Be aware that obesity, particularly in its extreme, is a risk factor for other illnesses including hypertension, diabetes, certain types of cancer, gall bladder disease, hyperlipidemia, and osteoarthritis. It may also be associated with a higher risk factor of coronary heart disease and postmenopausal breast cancer. Risk factors such as high serum cholesterol, elevated blood glucose, and high blood pressure can be curtailed by weight reduction in overweight adults. Lastly, be aware that increased energy expenditure is inversely associated with the risk of coronary heart disease and that leanness is associated with increased longevity.

People who are more than 30% over their ideal body weight are more likely to develop heart disease and stroke even if they have no other risk factors. Obesity is unhealthy because it increases the strain on the heart.

Adults with a history of teenage obesity are more likely to have CHD and athlerosclerosis as adults, assuming they remain obese. Males rather than females have an increased risk of dying from all causes (11).

Recent evidence indicates that how fat is distributed on the body may affect the risk of coronary heart disease. A waist:hip ratio greater than 1.0 for men indicates a significantly increased risk. For women, it is a ratio greater than 0.8. This means that a man's waist measurement should not exceed his hip measurement, and a woman's waist measurement should not be more than 80% of her hip measurement. Women tend to be shaped like pears and men like apples, with their protruding stomachs.

A University of Minnesota study of 41,837 Iowa women ages 55 to 69 indicated an apple shape—bigger waist than hips—was a better predictor of mortality than weight. For every 6-in. increase in waist measurement in a woman with 40-in. hips, the risk of death increased 60%. Lean women of apple shapes, like men, are at most risk of death. These investigators consider a waist : hip ratio of more than 0.75 to be apple shaped (12). The authors encouraged routine waist/hip ratio and circumference ratio measurements. Another University of Minnesota study investigated waist : hip ratios in overweight men and women as a function of weight losses. The ratios in men decreased more because they had greater decreases in the waist and smaller in the hips than women. Eliminating the weight loss factor, no evidence was found relating ratios to cardiovascular risk factors (13).

Weight distribution and breast cancer have been linked.

A University of Pennsylvania study on identical twins indicated that genes may be important in both weight and shape.

FAT LEVEL IN THE DIET
(PERCENTAGE OF CALORIES)

Over the decades of the 20th century, the level of fat (percentage of calories) in the diets of countries like the United States has risen to current levels variously estimated from 35 to 43%, most recently said to be 37%. Those levels of total fat, and particularly of saturated fatty acids of 16, 14, and 12 carbons constitute risk factors for cardiovascular diseases and certain other diseases including some forms of cancer.

A WHO study for the years 1986–1988 of 28 European and 5 other countries (United States included) highlighted a high rate of male mortality in Hungary and attributed this to >40% of total calories from fat, the fat being primarily lard, and a P : S (polyunsaturated : saturated) ratio of about 0.2 (14).

Nutritionists are now recommending a 30% level, or lower, of calories as fat with the intake of saturated fat not to exceed 10% of total energy. Most, but not all "experts" suggest the remaining 20% be divided equally between monounsaturated (10%) and polyunsaturated (10%) acids. Comparable reductions in total and saturated fat are recommended for children over 2 years of age, including adolescents. In September 1992, the American Academy of Pediatrics concurred with these recommendations (15).

Health and Welfare—Canada's Scientific Review Committee (16) has gone so far as to include recommendations on the omega (ω) acids. These are acids with double bonds (unsaturation), counting from the terminal methyl group. Specifically, Health and Welfare Canada said the omega-6 fatty acids should provide at least 3% of energy, and the omega-3 fatty acids at least 0.5% of energy, with an omega-6/omega-3 ratio in the range of 4 : 1 to 10 : 1. When the diet of infants contains no omega-3 C_{20} and C_{22} fatty acids (as from fish oils), 10% of the energy should be provided by alpha-linolenic acid, an omega-3 fatty acid. The FDA (United States Food and Drug Administration), which is working on standards for health claims allowed for fish oil, has cautioned manufacturers not to claim that these products are useful in preventing heart disease.

There is an expression—"moderation in all things." This holds true for what we eat; that is, keep your diet balanced. A mortality study on 4160 men and 6264 women ages 25–74 (mean ages men 53, women 48; 26% of the women and slightly greater than 26% of the men were 65 years of age or older) indicated that people who consume foods from each of the five food groups (grain, fruit, vegetable, meat, and dairy) had much lower relative risk of mortality than did people who omitted groups (17). Women who ate from all five weighed less. The data showed 46% ate no fruits, 25% no dairy, and 17% no vegetables. The author stated that renewed emphasis on consumption of a varied diet is not inconsistent with achieving the usual fat and fiber goals, and further investigation was recommended.

METABOLISM OF FATS

In the intestinal tract, dietary triglycerides (fats) are hydrolyzed to 2-monoglycerides and free fatty acids. These digestion products, together with bile salts, then form a micelle which moves to the epithelial cell membrane. There the fatty acids and the monoglycerides are absorbed into the cell and the bile acid is retained in the lumen. From 95 to 100% of most dietary fats are absorbed. In the intestinal wall, the monoglycerides and free fatty acids are recombined to form triglycerides. If the fatty acids have a chain length of 10 or fewer carbon atoms, as in the case of the medium chain length triglycerides, these acids are transported via the portal blood to the liver where they are metabolized rapidly. Triglycerides containing fatty acids having a chain length of more than 10 carbon atoms are transported via the lymphatics. These triglycerides, whether coming from the diet or from endogenous sources, are transported in the blood as lipoproteins. The triglycerides are stored in the adipose tissue until they are needed as a source of calories. The amount of fat stored depends on the caloric balance of the whole organism. Excess calories, regardless of whether they are in the form of fat, carbohydrate, or protein, are stored as fat. Consequently, appreciable amounts of dietary carbohydrate and some protein are converted to fat. The body can make saturated and monounsaturated fatty acids by modifying other fatty acids or by de novo synthesis from carbohydrate and protein. However, certain polyunsaturated fatty acids, such as linoleic or linolenic acids, cannot be made by the body and must be supplied in the diet (18).

Fat is mobilized from adipose tissue into the blood as free fatty acids. These form a complex with blood proteins and are distributed throughout the organism. The oxidation of free fatty acids is a major source of energy for the body. It has been known for many years that all of the usual dietary fats are of equal caloric value. The establishment of the common pathway of beta-oxidation by way of acetate, regardless of whether a fatty acid is saturated, monounsaturated, or polyunsaturated and whether the double bonds are cis or trans, explains this equivalence in caloric value (18).

Authorities for years have been saying metabolic rates vary between individuals. A May 1991 press release from the Beltsville USDA Energy and Protein Nutrition Laboratory contradicts this. More than 100 people were studied for 3 years in USDA's calorimeter, an unusual live-in laboratory designed to monitor food intake and measure human energy. To quote "Our tests found no metabolic magic; we've yet to see any evidence that there is any difference in people's efficiency in the way food is metabolized." The investigators also found that people have either faulty memories of their caloric intakes or they convince themselves that the number of calories eaten has no impact. These studies continue (19).

Another Beltsville USDA study, this from the Human Nutrition Research Center, will be of interest. Two hundred and sixty-six men and women were trained to record daily food intakes of conventional foods and their body weights were closely monitored. The workers concluded that the great majority in the United States underestimate their food intakes with those reporting the highest intakes tending to estimate their maintenance energy requirements more closely than those reporting lower intakes (20). The findings suggest caution in the interpretation of food-consumption data, which in turn can impact national survey data and then perhaps the nutritional status of the population.

PLASMA LIPIDS AND LIPOPROTEINS

Lipids are insoluble in water and circulate in plasma in association with certain specific proteins called apolipoproteins. Lipoproteins are large, macromolecular complexes of apolipoproteins and lipids in varying proportions. The four main classes of specific lipoproteins that circulate in plasma are chylomicrons, very low-density lipoproteins (VLDL), low-density lipoproteins (LDL), and high-density lipoproteins (HDL) (21).

Chylomicrons (composed of mainly triglycerides with cholesterol, phospholipids, and protein) are produced in the intestinal wall. When the chylomicrons enter the bloodstream they contact binding sites on capillaries where they release most of the triglycerides. The remainder of the chylomicron is absorbed by the liver. The liver produces a very low density protein (VLDL), the largest type of lipoprotein that contains added triglyceride. Like the chylomicrons, VLDL releases triglycerides at the capillaries. The remaining VLDL is called intermediate density protein, IDL. Some IDL is removed by the liver; the remainder is transformed to LDL (4).

The primary function of plasma lipoproteins is lipid transport. The major lipid transported in lipoproteins—triglyceride—is only slightly soluble in water, yet up to several hundred grams must be transported through the blood daily. Hence, transport mechanisms have evolved to permit the packaging of thousands of triglyceride molecules in individual lipoprotein particles, which deliver the transported lipid to specific cells. Fatty acids esterified to glycerol constitute approximately 90% of the mass and about 95% of the potential energy of the triglyceride molecule. Free fatty acids are transported in noncovalent linkage as albumin–fatty acid complexes. This latter mode of transport does not permit the high degree of selective targeting of fatty acids to specific sites that is permitted by transport in lipoproteins, but these two modes of fatty acid transport together provide a more versatile system for bulk movement of a major substrate for energy metabolism (21).

Cholesterol is the other major lipid transported in lipoproteins. It is not used for energy; it is the precursor of steroid hormones and bile acids and is a structural component of cellular membranes. In higher animals, including all mammals, it is transported mainly in the form of cholesteryl esters, which are synthesized in cells or in the plasma compartment itself. As with triglycerides, the transport of cholesteryl esters in lipoproteins permits specific targeting of cholesterol to tissues that require it for structural purposes or for making its metabolic products (21).

Two of the lipoprotein classes, chylomicrons and VLDL, are composed primarily of triglyceride. Chylomicrons transport exogenous (dietary) triglyceride, and VLDLs transport endogenous triglyceride. Chylomicrons are not normally present in postabsorptive plasma after an overnight fast. The VLDLs normally contain 10–15% of the plasma total cholesterol. LDLs contain cholesterol as their major component and normally contain most (60–80%) of the plasma cholesterol. HDLs are approximately half protein and half lipid and usually contain 20–30% of the total plasma cholesterol (21).

ROLE OF INGESTED FATTY ACIDS AND CHOLESTEROL IN CORONARY HEART DISEASE

Saturated Fatty Acids

A review of 420 dietary observations on 141 groups of subjects through 1991 concluded saturated acids increase serum cholesterol and are primary in that effect (22). The report also concluded that (1) not all saturated acids have the same effect; stearic acid has little or none; (2) monounsaturateds have no independent effect; and (3) polyunsaturates lower serum cholesterol. Both metabolic and field studies were involved.

Stearic acid continues to be considered not deleterious, but some investigators believe it may be thrombogenic. Use moderately C_{12} through C_{16} as furnished by coconut, palm, and palm kernel oils. Clinical, animal, and epidemiological studies demonstrate that increased intakes of these acids may increase the levels of serum total and LDL cholesterol and that these higher levels, in turn, may lead to atherosclerosis and increased risk of coronary heart disease (CHD). The supposed mechanism is the reduction in the uptake of LDL in the liver giving rise to an increased LDL level in the plasma.

Increasing evidence indicates that C_{16} or palmitic is the prime offender, particularly since its levels in oils and fats exceed those of C_{14} and C_{12}. Even here the picture is not consistent.

A recent study on 111 men and women as outpatients on a 7% saturated fat diet lowered LDL and HDL, both 5%. Cholesterol-lowering drugs were also studied and found to markedly reduce LDL while slightly raising HDL (23).

A recently reported 5-year study of non smoking women ages 30–84 showed a fivefold elevation of lung cancer risk on diets high in saturated acids (24).

There is evidence that the lower acids, below C_{12}, are metabolized promptly and probably have no effect on blood lipoprotein levels.

Monounsaturated Fatty Acids

Clinical studies indicate that the substitution of monounsaturated fatty acids for saturated fatty acids results in a reduction of serum total cholesterol and LDL cholesterol without a reduction in HDL-cholesterol (25). The monounsaturated fatty acids are, therefore, metabolically more favorable than saturated or polyunsaturated acids as high levels of the latter have been associated with adverse effects (recall that some of the polyunsaturateds are essential). The American Heart Association has suggested that the ratio of mono to poly should be 1.5 (26).

There is some evidence that oleic acid reduces HDL-cholesterol less frequently than does linoleic, which is another possible positive effect. In vitro analyses showed oleic to be protective of oxidative modification of lipoprotein. Oxidized LDL accelerates cell production leading to arterial injury. Monounsaturated acids, therefore, may protect and decrease LDL (27). Numerous recent studies are consistent relative to monounsaturated lowering LDL (28). In the last study, both mono and poly lowered LDL.

Polyunsaturated Fatty Acids

Clinical and animal studies provide firm evidence that omega or n − 6 (n minus six means position away from the terminal methyl group) polyunsaturated acids (linoleic and arachidonic) when substituted for saturated acids lower serum total and LDL cholesterol while usually causing some lowering of HDL-cholesterol. This finding may not be true for all levels of polyunsaturated fatty acids.

Fish oils containing large amounts of omega or n − 3 polyunsaturated acids (linolenic is the vegetable oil example) reduce plasma triglycerides and increase blood clotting time. Their effects on LDL-cholesterol vary, and long-term health effects of large doses are limited. These acids have also been implicated as having positive roles in vascular, inflammatory, and allergic diseases, and in platelet aggregation (clumping). Platelet clumping or aggregation in plaque can lead to thrombi which can lead to heart attacks and strokes. Limited epidemiological data suggest consumption of one or two servings per week of fish is associated with a lower CHD risk, but the evidence, so far, is insufficient.

Linoleic acid has been reported as an important risk factor in many chronic diseases, mainly through an adverse effect on the immune system. It was also said that the level of consumed linoleic is 10 times that needed for health and body maintenance.

Both N-6 and N-3 acids are important in prostaglandin biosynthesis. These are necessary hormone-like compounds that are mediators of the defense system. They are required especially for early development and growth of the fetus and infants. Recent data also suggests that these materials and their metabolites influence vision, mental development, and behavior characteristics of the child during critical periods of growth (29).

See the section on "essential fatty acids" for more information.

Trans Acids

Trans acids are unsaturated fatty acids with one or more double bonds in the trans position. In trans acids, the hydrogen atoms at the double bonds are on opposite sides of the chain. Cis acids have the hydrogen atoms on the same side, as are the long-chain radicals. Most trans acids consumed in the United States are formed by the hydrogenation process (see Chapters 3 and 6). It should be understood that this process is used for the production of both shortenings and oils and is performed to increase keeping quality or oxidative stability and to gain the desired levels of solids for appropriate plasticity. Some minor increases in saturated acids take place on partial hydrogenation.

Trans acids are naturally occurring. Tallow (beef fat) and butterfat contain levels of 2–7%. These result from biohydrogenation in the animals (ruminants). In the partial hydrogenation of vegetable oils, the trans acid levels range from about 10% to about 40%.

Most of the trans acids formed are of the monene or one double bond type. A common one is elaidic acid, 9-trans-C18:1 (hydrogenation also causes the double bond to move both directions on the hydrocarbon chain). Elaidic acid is the geometric or trans isomer to the very commonly occurring cis oleic acid.

Until recently it was believed that dietary trans acids did not influence serum cholesterol levels and posed neither a cardiovascular nor a carcinogenic risk. A recent clinical study (30) reported that the effect of trans fatty acids on the serum lipoprotein profile is at least as unfavorable as the other cholesterol-raising saturated fatty acids, because they not only raise LDL-cholesterol levels but also lower the HDL-cholesterol levels. This study was a short-term one and the high-trans products were not produced under typical U.S. hydrogenation conditions. The trans levels fed were higher than the calculated availability in the U.S. diet (31).

At a session on monounsaturates in nutrition at the 1991 annual AOCS meeting, several papers on trans fatty acids were presented. The data in those papers are consistent with the premise that stearic acid and trans monounsaturated acids have a similar influence on lipoprotein cholesterol levels (32).

Several additional clinical studies showed that trans acids raise total cholesterol, LDL, and Lp(a) [Lipoprotein (a)], but lower HDL. Still another study saw no apparent effect. In the last of the cited references, the authors concluded "the magnitude of lipid differences associated with trans fatty acids intake is sufficient to have important quantitative effects on the risk of myocardial infarction" (33).

The FDA has said "new data are rapidly emerging that imply that trans fatty acids raise LDL-cholesterol"; hence, it may consider trans acid labeling "in the near future" (34).

Dietary Cholesterol

Clinical, animal, and epidemiological studies generally indicate that dietary cholesterol raises total serum cholesterol and LDL-cholesterol levels and increases the risk of atherosclerosis (discussed in a later section) and coronary heart disease. There is substantial inter-individual and intra-individual variability in this response. High dietary cholesterol may contribute to the development of atherosclerosis and increased coronary heart disease risk.

Drugs are available that lower blood cholesterol levels. These should be used only when dietary changes will not suffice and when under the direction of competent medical advice. They should rarely be used by the elderly.

Gallstones, particularly in the United States, are made of cholesterol. Cholecystectomy, surgical removal of the gall bladder, is one of the most common operations in the United States. Obesity, rapid weight loss, and certain antilipemic drugs increase the risk of gallstones.

ESSENTIAL FATTY ACIDS

Linoleic and linolenic are considered to be the essential fatty acids, essential because (1) they cannot be synthesized by the body and, hence, must be supplied by the diet and (2) they are required for important bodily functions such as growth and good skin and hair qualities.

It has been estimated that of the total calories the minimum intake of linoleic acid should be about 3% with that of linolenic acid at about 0.5%. The current American diet is considered adequate to provide these levels. Investigations have shown that an intake linoleic of 3g/day is sufficient, whereas the current average intake is in the 20–25-g/day range: an excellent cushion. However, those excess levels give rise to more arachidonic acid that can possibly lead to hypersensitivity, depression of the immune system, and inflammatory diseases.

Intakes of linoleic, linolenic, monounsaturated fatty acids, and saturated fat not exceeding 5%, 3%, 10%, and 15% of the total fat consumption, respectively, have been recommended. These remarks are included to be examples of the disparity in recommendations (most recommend up to 10% of calories as saturated) and the complexity of the subject and the frequent lack of concensus.

Arachidonic (C_{20} 4 double bonds), eicosapentaenoic (C_{20} 5 double bonds), and docosahexaenoic (C_{22} 6 double bonds) have sometimes been referred to as essential. They are essential components of membranes such as those in the retina and central nervous system and are precursors of hormone like compounds, mediators in the defense system, referred to as prostaglandins. Because the body can synthesize these acids—linoleic acid the source of arachidonic and the other two from linolenic—none should be considered essential. Arachidonic is called an N-6 or omega acid and the other two N or N-3 acids where the numbers indicate the carbon atom locations of the first double bonds from the terminal methyl groups. The optimum ratio of N-6 to N-3 lies between 4:1 and 10:1.

Biological tissues and other functions are affected by eicosanoids (chemical modulators of the immune system). The response can include hemostasis, thrombosis, glucose homeostasis, and diabetes mellitus. Studies have shown that an excess of prostaglandin E_2 may cause immune depression. Releasing excess levels of arachidonic acid can cause problems with pathophysiological symptoms of disease following injury, infection, or inflamation. These might include inflamation, multiple organ failure, sepsis, and endotoxins (35).

In addition to the prostaglandins, the essential fatty acids are also precursors of additional biological active materials such as thromboxanes and leukotriexes.

As was mentioned under polyunsaturated acids, the essential fatty acids are crucial for normal fetal and child development. The human brain is large compared to body size and weight. Fetuses demand for essential fatty acids is greatest toward the end of term. Attention must be given to mothers' diets prior to conception, during pregnancy, and after, and to infant formulas, particularly for premature babies. This report also said that "maternal nutrition may not only affect the growth of the fetal and infant brain, but may have a long-term impact on the child's susceptibility to high blood pressure and heart disease in later life" (29). The entire part II of Issue 4 of *Nutritional Reviews* (1992) is probably worth perusing. It covers the 14th Marabou Symposium on the Nutritional Role of Fat. The 103rd Ross (Ross Laboratories) Conference on Pediatric Research in March 1992 in Australia was entitled "Lipids, Learning, and the Brain: Fats in Infant Formulas." All speakers reviewed and discussed all papers so that review of the material was very comprehensive. The report was published in June 1993. It contains a wealth of information (36).

LIPID PEROXIDATION (37)

Lipid oxidation occurring in the body is generally called peroxidation, whereas if the oxidation occurs in foods, it is generally called autoxidation (see Chapter 3). Both are considered deleterious. There is little question that autoxidation is deleterious since it gives rise to off quality (see also section on exposed fats in this chapter), particularly in flavor and its component aroma. Evidence continues to grow that lipid peroxidation is harmful to health. Related questions include: What levels of what oxidation products are adverse? What levels of what oxidation products will consumers note organoleptically and reject? It is believed that to some extent desirable flavors result from oxidation or oxidation products, as in fried snack foods. Chronic diseases and the process of aging have been found to be influenced by lipid oxidation. As has been reported (see previous section), arterial clogging in cardiovascular disease apparently begins when LDL is oxidized in a reaction induced by free radicals having their source in oxidized polyunsaturated acids. Lipid oxidation products may accelerate all three phases of chronic heart disease: initiation, progression, and termination. Interestingly, there is some evidence that lipid oxidation might help destroy cancer cells.

Fortunately, many consumers will react negatively to the adverse quality oxidized lipids give to products. But what is the margin of safety? What are the "facts?" What education is needed? This is undoubtedly a potential global problem, not just for the United States or the developed world.

Methods for measuring the oxidative status of lipids and lipid stability are receiving increased attention. Approaches include front face ratiometric fluorometry, potentiometry high-performance liquid chromatography, spectrophotometry, nuclear magnetic resonance spectroscopy, headspace gas chromatography of volatile oxidation products, chemiluminescence assays, and conductimetric determination. The last approach has resulted in an official method adopted by the Japan Oil Chemists' Society as an alternate to the active oxygen method (AOM). (See Chapter 14).

ANTIOXIDANT NUTRIENTS

It is generally believed that oxidized LDL causes buildup of fatty plaque on arterial walls. Interest has increased in the past few years in the free radical theory of disease causation, particularly in vascular diseases and certain forms of cancer. These developments have led to the investigation of dietary agents, the antioxidant nutrients (mainly vitamins A, C, and E), in a possible prophylactic, even curative, role in the disease process.

Vitamin E is the dietary antioxidant that has received the most attention. Two recent studies, one on 87,245 nurses (38) and one on 45,720 male health professionals (39) on vitamin E supplements for several years, reported decreased incidences of heart attacks of 46 and 26%, respectively. The optimum dosage in both studies was said to be 100–250 IU, several times the recommended RDA. The authors indicate the evidence is not strong enough to prove a cause-and-effect relationship and that a clinical trial is indicated. The question of possible long-range contraindications of continued ingestion of high levels of E would also need to be answered. Interestingly, megadoses, greater than 400 units, were not effective.

Vegetable oils are excellent sources of Vitamin E (tocopherols). Corn, cottonseed, peanut, safflower, sunflower, and soybean all run close to 0.1% (0.07–0.1%).

Studies on Vitamin C, another natural antioxidant, have yielded mixed results; some positive, some no effect.

It has been intimated that oxidized LDL attaches more readily to the arterial wall. A laboratory test showed that oxidized LDL is toxic and kills the artery cells. Repair of the wound leads to the buildup of tissue.

Closely related to this probable benefit of natural antioxidants is their role in "controlling" free radicals as they may lead to pathological effects such as vascular diseases and cancer. A free radical is defined as any chemical species that has one or more unpaired electrons. This results in very reactive compounds. Oxidation is a natural and needed reaction in metabolism. Highly reactive hydroxyl radicals, OH^-, result. These can attack DNA, protein, and polyunsaturated fatty acid residues of membrane phospholipids, among others. With the latter, a peroxyl radical is formed. Vitamin E quenches this radical. If the supply of vitamin E is inadequate, a chain reaction takes place that may lead to damaged tissue. The evidence in the literature begins to make an overwhelming case for the existence of a relationship between high blood levels of antioxidant nutrients and a lowered incidence of disease (40).

An international meeting on antioxidants, free radicals, and polyunsaturated fatty acids held in May 1993 in Lystrup (Copenhagen, Denmark) was reported on recently in *Inform* (41). Selenium was mentioned as having a physiological and clinical role in antioxidant protection.

It has also been said that vitamin E helps reduce muscle damage from oxygen free radicals produced during exercise (42).

Experts, in general, caution for more study (43).

Time magazine closed a recent (5/31/93) health article by saying, "Unfortunately for a quick-fix society, getting plenty of exercise, cutting down on dietary fat and quitting smoking remain far better prescriptions for preventing heart disease than anything one can obtain from a vitamin bottle." Maybe "designer" foods will force *Time* to change its mind.

CARDIOVASCULAR DISEASES

The section on leading causes of death stated that cardiovascular diseases, primarily heart attacks and stroke, are the leading causes of death in the United States. Atherosclerosis (next section) is the major cause of cardiovascular diseases. Its cause, however, is not direct but via the released thrombi or blood clots from the formed fatty plaque.

Cardiovascular diseases are chronic degenerative diseases of complex etiology that often are associated with aging. A number of risk factors have been identified from epidemiological studies. These include positive family history of cardiovascular diseases, cigarette smoking, hypertension (high blood pressure), elevated serum cholesterol, obesity, diabetes, physical inactivity, male sex, and excessive stress. Although these risk factors have been associated statistically with the incidence and mortality of cardiovascular disease, cause-and-effect relationships are lacking.

The 1993 American Heart Association's "Heart and Stroke Facts" states the four major modifiable risk factors are high blood pressure, high blood cholesterol, cigarette smoke, and physical inactivity. The last was added in 1993.

The advisory board to the International Health Conference held in 1992 noted that cardiovascular disease is largely preventable (44). It said, "Most cardiovascular disease is brought about by some combination of smoking, high blood pressure, elevated blood cholesterol, unhealthy dietary habits—including excessive alcohol consumption—obesity, a sedentary lifestyle, and psycho-social stress." It further noted that reducing or eradicating such risk factors "will produce declines not only in cardiovascular disease but also in other noncommunicable diseases which share similar risk factors—for example, many types of common cancer and lung and liver disease."

The amount of fat ingested and its types are vitally important.

Baldness in men has been suggested as a predictor of CHD. A recent study (1993) tends to confirm this hypothesis (45); 1437 men were studied: 665 upon having their first nonfatal myocardial infarction and 772 controls. The correlation is with vertex balding, or loss on the top of the head, not receding hair lines.

The investigators emphasized that any risk linked to baldness is smaller than for smoking, hypertension, and abdominal obesity.

ATHEROSCLEROSIS

Arteriosclerosis is a general term for the thickening and hardening of arteries. Some hardening of arteries normally occurs when people grow older. It is characterized by deposits of fatty substances, cholesterol, cellular waste products, calcium, and fibrin (a clotting material in the blood) in the inner lining of an artery. The resulting buildup is called a plaque.

Plaques may partially or totally block the blood's flow through an artery. Two things that can happen where plaque occurs: (1) bleeding (hemorrhage) into the plaque or (2) formation of a blood clot (thrombus) on the plaque's surface. If either of these occurs and blocks the entire artery, a heart attack or stroke may result (4).

Atherosclerosis affects large and medium-sized arteries. The type of artery and where the plaque develops varies with the individual.

Atherosclerosis is a slow, progressive disease that may start in childhood. In some people, this disease progresses rapidly in their third decade; in others, it does not become threatening until they are in their fifties or sixties. It is rare in children, and no evidence has been seen with children under 2 years of age.

Atherosclerosis is a complex process. Precisely how it begins or what causes it is not known, although several theories have been proposed.

Many scientists believe atherosclerosis begins because the innermost layer of the artery (endothelium) becomes damaged, and, over time, fats, cholesterol, fibrin, platelets, cellular debris, and calcium are deposited in the artery wall. Gradually, these substances build up and eventually narrow and block the artery, similar to scale forming on the insides of pipes.

Three of the possible causes of damage to the arterial wall are (1) elevated levels of cholesterol and triglyceride in the blood, (2) high blood pressure, and (3) cigarette smoke. Cigarette smoke is particularly potent in aggravating and accelerating the development of atherosclerosis in the coronary arteries, aorta, and arteries of the legs.

Once cells in the artery wall are damaged, they may separate from the wall, exposing the tissue (collagen, smooth muscle, and other tissue) underneath. Initially, platelets (elements in the blood that help form blood clots) stick to the collagen, which ultimately leads to the formation of plaque. Over time, the extent of atherosclerosis increases, reducing the diameter of the artery. Eventually a blood clot may form at the site of damage, blocking the artery and stopping the normal flow of blood.

Scientists are studying other ways in which platelets may play a role in atherosclerosis. For example, platelets are involved in forming a group of substances called prostaglandins, one of which may damage arteries. Platelets also contain a substance called "platelet growth factor," which can stimulate the growth of smooth muscle cells. These cells are normally present in the artery wall, but their abnormal growth and proliferation is believed to be one of the earliest events in developing atherosclerosis.

One of the more recent theories suggests that plasma lipoproteins are trapped within the artery wall. When this happens and they increasingly accumulate, they oxidize, which leads to "modified" lipoproteins that are rapidly taken up by smooth muscle cells. This, in turn, leads to the formation of foam cells and the development of a fatty streak.

In 1934, it was noted that atherosclerosis and atherosclerotic diseases in many parts of the world corresponded to the overconsumption of fats and cholesterol.

Early in the 1950s, the serum-cholesterol-lowering effects of the polyunsaturated acids were discovered, and epidemiological and human experimental studies were made on this issue. The role of dietary cholesterol remained uncertain until the 1960s when several careful experiments in humans showed that there was a modest but definite effect.

Atherosclerosis may well be reversible.

CHOLESTEROL

Cholesterol is a soft, fatlike substance found in all of the body's cells. It is used to form cell membranes, certain hormones, and other necessary substances.

People obtain cholesterol in two ways. The body, primarily the liver, produces varying amounts, usually about 1000 mg a day. An additional 400–500 mg or more can come directly from foods. Foods from animals, especially egg yolks, meat, fish, poultry, and whole milk dairy products, contain it. Foods from plants do not. Typically, the body makes all the cholesterol it needs, so it is not something that people have to consume to maintain their health.

Besides being present in human tissues, cholesterol is also found in the bloodstream. The blood transports it to and from the various parts of the body via lipoproteins.

Levels of total serum cholesterol below 200 mg/dl* are considered desirable, 200–239 of concern, and 240 and above at risk. Be aware that other factors, including genetics, influence the levels.

The earlier reported review (22) of 141 groups of subjects through 1991 had this to say about cholesterol:

- Dietary cholesterol increases levels, hence needs to be considered.
- Limited data showed LDL changes paralleling serum cholesterol.

---

*dl = deciliter; terminology had been 100 ml. They are one and the same.

Behavior of HDL could not be satisfactorily predicted. Be aware that this extensive review excluded some important considerations such as trans/hydrogenated fats and fish oils.

Clinical, animal, and epidemiological studies generally indicate that dietary cholesterol raises serum total cholesterol and LDL cholesterol levels and increases the risk of atherosclerosis and coronary heart disease. As indicated, there are substantial inter-individual and intra-individual variables in this response. High dietary cholesterol may contribute to atherosclerosis and coronary heart disease risk.

Evidence is accumulating that indicates lowering cholesterol has a beneficial effect on heart disease. Medical doctors have suspected since the 1950s that aggressive people (Type A) are 50% more likely to suffer heart attacks than nonaggressive (Type B). This may be due to lower HDLs in Type A. A recent study showed that Type A HDLs were about 10 points lower than Type Bs. A one point lowering in HDL is thought to increase the risk of heart attacks by about 3% (46).

Several studies released in 1992 and based on data from "hundreds of thousands of people" suggest there is a risk to having too low a cholesterol level, 160 or less. These individuals die from other causes at a rate equal to those with high cholesterol levels. The researchers stated that these new findings do not question the standard advice of the use of diets to lower cholesterol when levels are high.

A mid-1992 article indicated it is time to change directions on the health policy on blood cholesterol (47). The three conclusions were that (1) there is an association between blood cholesterol and non-cardiovascular deaths in men and women; (2) there is *no* association between high cholesterol in deaths of women; and (3) there is an increase in non-CHD death rates that is similar in magnitude to the decrease in CHD death rates.

Two related studies appearing in mid-June in *Journal of the American Medical Association* (in 1993) indicate that the U.S. adult cholesterol level now averages 205, having dropped 6–8% in the past 30 years. The figure may go down further. The public health programs are working. Dietary intervention is paying off. The downward trend has coincided with a declining CHD mortality. Whereas 26% of adults had levels above 240 in 1975, now only 20% do. In addition to dietary changes, increased exercise, fewer smokers, and cholesterol-lowering drugs contribute (48).

Physicians should be cautious about initiating cholesterol-lowering treatment in men and women above about 65–70 years of age.

Data from the Framingham study indicate there is, relative to mortality risk,

- Positive relationship of cholesterol at age 40
- Positive but not significant or negligible relationship at ages 50–70
- Negative relationship at age 80 (49)

The reader needs to be aware that in spite of the numerous promising reports, there is contrary evidence. All is not clear.

- In women, cholesterol levels appeared to have no influence on how they died from all causes. For those with high cholesterols, the higher death rate due to heart attacks was balanced by lower rates from certain strokes and other diseases. In men, the only groups that died more often were those whose cholesterol levels were either very high or very low. Men with low levels were only slightly less likely to die than men with levels > 240 (50).
- It has been recommended by the National Education Program Adult Treatment Panel that all adults over age 20 know their cholesterol levels. This may not be justified and even be unwise. It has been concluded that, (1) there is no justification for routine cholesterol tests for men under 35 and women under 45, and (2) early screening may lead young adults to embark on a regimen of drug therapy whose long-range effects are unknown. They can wait until they get older to diet or start drug treatment (51).

The relationship between total cholesterol levels and the prevalence and incidence of, as well as mortality from, atherosclerotic coronary heart disease has been confirmed by autopsy studies.

LDL-Cholesterol

LDL is the major cholesterol carrier in the blood; about 60–80% of the plasma cholesterol is carried by LDL. Some of this cholesterol is used by the tissues to build cells and some is returned to the liver, but if there is too much LDL cholesterol circulating in the blood, cholesterol may also be deposited in artery walls causing the plaques and atherosclerosis. It is called the "bad" cholesterol because of these deposits and because lower levels of LDL reflect a reduced risk of heart disease.

LDL levels of greater than about 160 mg/dl are generally considered to indicate increased risk.

Dietary cholesterol has the greatest influence on LDL. On average, 100 mg of dietary cholesterol per kcal elevates LDL by 8–10 mg/dl. The usual range of human intake of dietary cholesterol is 300–600 mg/day or about 100–300 mg/1000 kcal.

HDL-Cholesterol

The high-density lipoprotein is produced primarily in the liver and intestines and released into the blood stream. As VLDL (see plasma/lipids section) and chylomicron particles release triglycerides into the body's cells, fragments containing proteins, fats, and cholesterol break away. It is thought that HDL picks up the cholesterol and brings it back to the liver for reprocessing or excretion. Some researchers believe HDL may also remove excess cholesterol from fatsated cells, possibly even those in artery walls.

HDL cholesterol is considered the "good" cholesterol because it clears cholesterol out of the system and high levels of it are associated with a decreased risk of heart disease.

The levels of HDL and LDL in the blood are used to evaluate the risk of atherosclerosis.

Generally speaking, levels below about 40 mg/dl of HDL cholesterol are considered to indicate a risk situation. One investigator has obtained evidence of what might seem to be obvious. Low HDL levels are caused by a combination of low rate of synthesis and a rapid rate of breakdown. The studies indicate more rapid breakdown is associated with obesity, lack of exercise, smoking, high triglycerides (above about 200 mg/dl), and the use of muscle-building steroids. It was claimed that the desirable HDL level for men is above 45 with a high-risk level below 35, and for women, desirable above 60, and high risk below 45. Further it has stated the ratio of total cholesterol to HDL is important; desirable below 4.0, borderline 4.0–6.0, and high risk above 6.0.

As a rule, women have higher HDL levels than men. The female sex hormone estrogen tends to raise HDL, which may help to explain why premenopausal women are usually protected from developing heart disease. Estrogen production is highest during the childbearing years.

Low levels of HDL do not necessarily represent a risk factor. It has been said that individuals with "alarmingly low" levels will present only a normal risk.

Lipoprotein (a)

Lipoprotein (a) [Lp(a)] is an unusual blood lipoprotein. It is a transporter of cholesterol that can bind with blood clots and raise the risk of a heart attack. Its existence has been known since the early 1960s but recently has received additional attention and renewed study (52). It may be an explanation for previously unattributable risks for coronary heart disease: that is, its blood levels are high in individuals whose vulnerability cannot be laid to other obvious causes.

Other interesting facets of Lp(a) include the following:

• A portion of its molecule is identical to LDL.

• There are no accepted ways to lower high plasma levels. It has been said that dietary change and other measures have no effect that neither foods nor drugs can control.

• It is reputed to be an inherited factor; an estimated 25% of males under 60 have inherited high concentrations.

• The level in the blood varies almost 1000-fold.

• The level tends to be stable throughout lifetimes.

• It is found in atherosclerotic plaques.

• At this time there is limited testing capability.

Many studies have shown high levels in the blood are associated with heart attacks, strokes, narrowing of arteries, and closure of vessels.

A recent study of 27 mildly hypercholesterolemic men for eleven weeks on a high-trans (elaidic) diet led to significant elevations compared to high butterfat, oleic rich, and palmitic rich diets (53).

TRIGLYCERIDES

Triglyceride levels normally range from about 50 to 250 mg/dl, depending on age and sex. As people get older and sometimes heavier with age, their triglyceride and cholesterol levels tend to rise. Women also tend to have higher triglyceride levels than men (4). An elevated blood triglyceride level and lower HDL is often accompanied by an increase in LDL and total cholesterol.

Several clinical studies have shown that an unusually large number of people with coronary heart disease also have high levels of triglycerides in the blood (hypertriglyceridemia). However, some people with this problem seem remarkably free from atherosclerosis. Thus, elevated triglycerides, which are often measured along with HDL and LDL, may not directly cause atherosclerosis but may accompany other abnormalities that speed its development.

A fairly recent foreign study (early 1992) Associated Press news release said "blood fats called highest heart risks." Individuals whose total level of blood cholesterol contained a high proportion of LDL and high triglycerides were 3.8 times as likely as others to have a heart attack. This was said to be true even in individuals whose total cholesterol is normal or only slightly elevated. The study also showed that heart attack risk can be cut 71% by lowering triglycerides and raising HDL. This can be done by losing weight and exercise, or by drugs.(54)

CANCER

As was stated in the section on leading causes of death, cancer is the number two killer in the United States.

Cancer is considered more frightening than cardiovascular disease. Probably the two key factors are (1) cancer can be very debilitating and lingering, and (2) it kills more children, ages 3–14, than any other cause.

What are some of the incidences, in estimated cases for 1993? In parentheses are the 1991 figures. Note most have increased. Skin cancer leads the list at over 700,000 (600,000) (5), followed by breast at 182,000 (175,000), lung at 170,000 (161,000), colon and rectum at 152,000 (157,500), prostate at 165,000 (122,000), bladder at 52,300 (50,200), uterus and cervex at 44,500 (46,000), oral at 29,800 (30,800), pancreatic at 27,700 (28,200), leukemia at 29,300 (28,000) and ovarian at 22,000 (20,700). Of these, high fat is considered by some health professionals to be a significant causative factor in colon/rectum cancer and a possible cause for pancreatic cancer. Dietary fat may be a factor in breast and prostate cancers, and obese individuals are at increased risks for breast, colon/rectum, and uterine cancers.

Of the estimated 9100 (8,500) deaths from skin cancer in 1993, 6800 (6500) are expected from malignant melanomas.

Normally, the cells that make up the body reproduce in an orderly manner so that worn-out tissues are replaced, injuries are repaired, and growth of the body proceeds. Occasionally, certain cells undergo a change that is abnormal and, thus, begin a process of uncontrolled growth and spread. These cells may grow into masses of tissue called tumors. Some tumors are benign (noncancerous) and others are malignant (cancerous). The danger of cancer is that it invades and destroys normal tissue. In the beginning, cancer cells usually remain at their original site, and the cancer is said to be localized. Later, cancer cells may metastasize (that is, they invade distant or neighboring organs or tissue). This occurs either by direct extension of growth, or by cells becoming detached and carried through the lymph or blood systems to other parts of the body. Metastasis may be regional—confined to one region of the body— where cells are trapped by lymph nodes which are omnipresent. If left untreated, however, the cancer is likely to spread throughout the body. That condition is known as advanced cancer, and usually results in death. Systems have been designated to define what stage a cancer has reached. Because cancer becomes more serious with each stage, it is important to detect cancer as early as possible. Aids to early detection include knowing cancer's seven warning signals and the risk factors. The former are (1) change in bowel or bladder habits, (2) a sore that does not heal, (3) unusual bleeding or discharge, (4) thickening or lump in breast or elsewhere, (5) indigestion or difficulty in swallowing, (6) obvious change in a wart or mole; and (7) nagging cough or hoarseness. In addition to the fat risk factors listed earlier, one should avoid getting old, avoid smoking, avoid excessive sunlight, select the proper parents, select the appropriate sexual practices, and eat fiber.

The mechanisms by which dietary fat may influence tumor promotion remain obscure. Hypotheses that have been suggested include alterations in cellular permeability, effects on the host immune system, or perturbation of prostaglandin metabolism. Some epidemiological correlations between fat consumption and colon cancer have led some investigators to suggest that fat may enhance the development of intestinal tumors by stimulating production of bile acids, some of which may act as promoters of tumorigeneses. Some studies also have associated increased colon cancer incidence with low serum cholesterol levels. In all cases, more work is needed to determine if those associations have any health risk.

The National Research Council's Committee on Diet, Nutrition, and Cancer (1982) concluded

> that of all the dietary components it studied, the combined epidemiological and experimental evidence is most suggestive for a causal relationship between fat intake and the occurrence of cancer. Both epidemiological studies and experiments in animals provide convincing evidence that increasing the intake of

total fat increases the incidence of cancer at certain sites, particularly the breast and colon, and, conversely, that the risk is lower with lower intakes of fat. Data from studies in animals suggest that when fat intake is low, polyunsaturated fats are more effective than saturated fats in enhancing tumorigenesis, whereas the data on humans do not permit a clear distinction to be made between the effects of different components of fat. In general, however, the evidence from epidemiological and laboratory studies is consistent.(21)

The National Research Council's (NRC) Committee on Diet and Health (1989) further concluded

that the relationship between dietary cholesterol and cancer is not clear. Many studies of serum cholesterol levels and cancer mortality in human populations have demonstrated an inverse correlation with colon cancer among men, but the evidence is not conclusive. Data on cholesterol and cancer risk from studies in animals are too limited to permit any inferences to be drawn.(21)

The conclusion on fat was based largely on the consistency between epidemiologic and animal evidence. International correlation studies show direct associations between per-capita availability of dietary fat and the incidence of mortality from cancer at such sites as the breast, prostate, and gastrointestinal tract. Further evidence is provided by observational epidemiologic studies and by experimental results showing that animals given certain carcinogens and then fed high-fat diets develop cancers of the mammary gland, pancreas, and intestinal tract more readily than do animals on low-fat diets.

Epidemiologic data from different countries show that cancer incidence and mortality correlate more positively with total dietary fat than with any one type of dietary fat. On the other hand, results of experiments in animals demonstrate that vegetable oils containing omega-6 PUFAs (Polyunsaturated Fatty Acids) promote carcinogenesis more effectively than SFAs, whereas fish oils containing omega-3 PUFAs tend to inhibit carcinogenesis. These findings are not necessarily in conflict with the epidemiologic data, as the studies were conducted primarily with individual fats and oils with fatty acid compositions that are quite different from those of the mixtures of fats and oils present in the diets of humans.

Studies have shown a strong positive correlation for total fat and animal fat intake and breast, colon, and, to a lesser extent, prostate cancer (27). A 4-year study in 20 countries yielded results strongly associating total fat intake with breast, colon, and prostate cancers after adjustments made for total caloric intake. Monounsaturated acids had no positive association with cancer at any site. However, saturated and polyunsaturated acids were positively associated with incidences of cancer of the breast and colon. A study at the Centers for Disease Control indicated a diet high in fat significantly increased the risk of colon cancer. The link between diet and breast cancer is tenuous.

A recent animal study involving four diets (55) supports the contention that total calories are most important rather than the portion contributed by fat. This same study concluded low fat and high fiber may not reduce the incidence of breast cancer in middle-aged women.

A 10-year effort by the American Cancer Society on over 750,000 men and women in the United States gave evidence for colon cancer reduction by the consumption of more fruits and vegetables (56).

A fairly recent report, April 1993 (57), stated that the high rates of breast and prostate cancers in the United States may be due to low soybean product consumption rather than high fat. A University of Illinois study also cited in Ref. 57 showed an average 12% drop in blood cholesterol in diets fortified with soy ingredients.

The evidence suggests that the risk of breast cancer and, to a possibly greater extent, of colon, prostate, and ovarian cancers, is associated with dietary fat, total or saturated fatty acids. What is considered more important is total calorie intake. This has been known to be true for animals for over 50 years and recent studies present mounting evidence for humans.

No link, no adverse influence was found in middle-aged women between fat intake and breast cancer. The study involved 89,494 registered nurses, ages 34–59, over a 9-year period via questionnaires. The average fat intake was 31–47% (58). The investigators encourage women on a low-fat diet to stay on it to help avoid colon cancer and heart disease.

Epidemiological studies have long suggested a link between high-fat diets and certain cancers of the female reproductive system.

According to one investigator dietary fat may be a modulator of carcinogenesis rather than a promotor; that is, it may enhance or inhibit depending on the "experimental design" (2). This same individual has stated that only linoleic acid has been shown "clearly and unequivocally to enhance tumor growth when fed to rodents" when on an excessive amount of calories.

Another worker has said that "an important unresolved issue in the diet and cancer area is whether the best diets to reduce heart disease risk may increase breast cancer risk in women and colon cancer risk in men."

The last individual report to be mentioned speaks to the observation that ether lipids can be antitumor drugs. 1-octadecyl-2 methylglycerocholine reduced tumor growth in animals, and hexadecyl-phosphocholine, which is less toxic, has been used in clinical trials on patients with skin metastases (59).

STROKE

Stroke is second to heart attacks in incidence and cause of death insofar as cardiovascular disease is concerned.

Stroke is the form of the disease that affects the central nervous system. A stroke occurs when a blood vessel serving the brain bursts or is clogged by a blood clot or some other particle. That portion of the brain which is affected dies in a matter of minutes due to the lack of oxygen, and that portion of the body served by those brain cells is also unable to function. These effects are often permanent.

Of the four main types of stroke, two are caused by clots. If the clot, referred to as a thrombus, has its origin in the vessels in the brain, the stroke is called a cerebral thrombosis; if the clot, referred to as an embolus, has its origin elsewhere and is carried to the brain, it is called a cerebral embolism. These two constitute 70–80% of all strokes. Recall that the origin of the clots is usually from arteries damaged by atherosclerosis.

Risk factors that can be treated are high blood pressure, heart disease, cigarette smoking, high red blood cell count, and transient ischemic attacks.

Uncontrollable risk factors include heredity, sex, age, and race. Factors that indirectly increase stroke risk include elevated blood cholesterol and lipids, and obesity.

HYPERTENSION (HIGH BLOOD PRESSURE)

Blood pressure is a result of two forces—one created by the heart as a pump and the other by the arterial vessels as they resist the blood flow.

In 90–95% of the cases of high blood pressure, the cause is unknown. These cases are usually effectively treated by drugs. In the remaining cases, high blood pressure is a symptom of a recognizable problem such as a kidney abnormality, a tumor of the adrenal gland, or a congenital defect of the aorta. Correcting the root cause usually returns the pressure to normal.

There is no "ideal" blood pressure reading. A typical pressure for an adult might be 127/78 mm of Hg. For most adults, a blood pressure of less than 140/90 is reasonable; that is, there is no cause for alarm. Hypertension exists when the systolic ("numerator") pressure is equal to or greater than 140 and/or the diastolic ("denominator") pressure is equal to or greater than 90 for an extended period of time.

What about fats? Existing evidence indicates that total dietary fat has little effect, if any, on high blood pressure. However, in hypertensive individuals, in a few but not all studies, increasing the dietary P:S ratio (polyunsaturates:saturates) has resulted in a modest decrease in pressure. Studies on monounsaturated suggest no effect.

Lowered blood pressure can be harmful. It may be that slightly elevated blood pressures (diastolic) in the elderly are actually optimal (60). It was found that risks of irregular heart beats increased when pressures fell below 85, especially if the hearts were thickened from hypertension. Abnormal beats were seen in about 50% of the patients with readings below 85 compared to 10% of those with readings of 85–94.

Preliminary findings from a Johns Hopkins study suggest postponement or avoidance of hypertension can be obtained by losing a little weight and using less salt (sodium). Just over 2 mm lowering of pressure would cut the nation's stroke rates by 6% and CHD by 4%.

Ingestion of ascorbic acid (Vitamin C, an antioxidant) by 168 healthy individuals (103 women and 60 men) on regular diets lowered blood pressure, both systolic and diastolic (61). The authors speculate as to ascorbate's role. The results agree with previous reports including 7 studies in the United States, Finland, and Japan involving over 12,000 subjects (62).

Recent studies in Oregon suggest that for a number of people it is not too much salt but too little calcium.

CLOTTING (HEMOSTASIS)

There have been several reports that eicosapentaenoic acid (one of the N-3 fatty acids from many fish oils) forms a unique type of phospholipid in human platelets after consumption of linolenic acid (which converts to the $C_{20:5}$ acid) that may reduce platelet aggregation, thus the clotting tendency. Some reports have suggested that saturated fatty acids are prothrombogenic compared to polyunsaturated fatty acids and that omega-3 polyunsaturated fatty acids are antithrombogenic compared to omega-6 polyunsaturated fatty acids.

A recent workshop on Dietary Fatty-Acids and Thrombosis (63) pointed out that improved methodology is needed to measure platelet aggregation. A conclusion reached was that there is no direct evidence that dietary long-chain saturated fatty acids are thrombogenic in humans. The FDA has been concerned about the type of information that is required for predicting thrombogenic risk of diets relative to nutrition, labeling, and health messages for foods. The proceedings of the cited workshop were published as a supplement to an *American Journal of Clinical Nutrition*, October, 1992 issue (64). Another conclusion from that workshop was that the role of oxidative processes in atherogenesis and thrombosis is unclear.

OTHER POSSIBLE DISEASE IMPLICATIONS

Mention has been made in the literature of dietary fats being implicated in gallbladder disease, diabetes including diabetic neuropathy, skin problems including psoriasis, arthritis including rheumatoid, retinal defects, multiple sclerosis, and other autoimmune, inflammatory, and cell-vessel-wall problems.

"EXPOSED" FATS

Chapter 3 discussed the common chemical reactions fats may undergo. Three of these may take place under conditions of exposure to heat, light, oxygen, and others and/or use giving rise to reaction products that may adversely influence quality, both organoleptic and nutritional. The reactions are hydrolysis, oxidation, and polymerization.

The word "exposed" is used here to include the terms "oxidized" fats and "heated" fats. The conditions of exposure or the reaction conditions vary for these two types; hence, the nature, types, and levels of reaction products will differ. Neither type, per se, has aroused much interest on the part of nutritionists in recent years as compared to concerns with the constituent unoxidized fatty acids, for example. Lipid peroxidation, that is, oxidation of lipids in the body, is an entirely different matter (see an earlier section).

Oxidized lipids are formed either spontaneously or by reaction with light. Large intakes (by test animals) have resulted in a number of adverse symptoms, including reduced growth, diarrhea, gastrointestinal bleeding, increased rates of abortion, and so on. The poisonous substances were stated to be malonic dialdehyde and 4-hydroxynoneal, via interaction with cell membranes. In view of the likely off flavor of such exposed materials, it is doubtful that they would be ingested in quantities that would cause problems. This is not known, however.

Heated fats, not surprisingly, have received more attention. These are ingested via fried, baked, and other finished food preparations.

For details, the reader is referred to a review of a symposium on "The Chemistry and Technology of Deep Fat Frying" held by the Institute of Food Technologists at its June 1990 annual meeting. The overview of this effort has been published (64). The authors of the review article on "Safety Aspects of Frying Fats and Oils" presented at this symposium concluded that "although excessive cooking conditions can produce toxic compounds in fats and oils, the fats and oils become unacceptable from a sensory standpoint sooner." As the authors pointed out, the desirable and appreciated sensory qualities of fried foods are destroyed by excessive cooking. In a table in the article summarizing animal feeding studies on commercially heated fats, no adverse effects were, by far, the most frequent observation. Reduced growth rate was noted in two of the nine reported studies. In a table on overheated fats, the adverse physiological conditions noted in the test animals included weight loss, loss of hair, diarrhea, and enlarged livers. The authors also stated that heating of fats can result in antinutritional compounds that may be enzyme inhibitors, vitamin destroyers, gastrointestinal inhibitors, and/or potential mutagens. They recommended that the intake of fried foods be moderated as part of a balanced diet.

CANOLA OIL

Canola oil is an excellent example of the successful application of biotechnology (see the next section) to the composition of fats. It is derived from a specific type of rapeseed and is known as low erucic acid rapeseed oil.

Historically, rapeseed oil contained about 50% or more of erucic acid, a C_{22} monounsaturated acid, with lesser amounts of oleic, linoleic, linolenic, palmitic, and stearic acids. It was little used in the Western world because of safety concerns. Fatty infiltration of the heart and necrotic lesions have shown up in animal feeding studies.

Selective plant breeding in Canada, starting in the 1970s changed the composition radically. Erucic acid now runs no more than 2%; oleic is quite high, up to as much as 62%; significant quantities of linoleic (about 20%) and linolenic (about 10%) can be present; and saturated fatty acid levels are low (about 6–8%), making it a nutritionally desirable oil. It is now by far the major edible oil used in Canada, and use in the United States is increasing. Interestingly, the early rapeseed varieties were grown in Canada in the 1940s as an alternative lubricating oil for the British Navy.

In 1985, canola oil was approved by FDA (maximum of 2% for erucic acid) and the initial consumer product was on the market shortly thereafter.

BIOTECHNOLOGY

Biotechnology makes use of biological systems. It has wide applicability in food science. Classical methods of biotechnology can be quite old, for example, fermentations. New developments apply DNA techniques.

It is now possible to manipulate genes in plants, animals, and microbes to produce new and unique products. Basic research probing the molecular biology and biochemistry of enzymes and the molecular biology, biochemistry, metabolism, and physiology of microbes, plants, and animals is necessary to provide the desired solutions. The potential gains are enormous (65). The cited article lists prioritized research goals and their estimated costs for the next few years.

The previous section lists the best-known plant breeding or genetics effort to alter, in a controlled fashion, an oil or fat. This is but one example of this type of biotechnology. Others, in recent years, have raised the oleic acid levels from about 30% to 80% or more in oils such as safflower and sunflower.

Herbicide tolerances and insect and virus resistances have been inserted into a number of important crops by recombinant DNA technology (66).

Lipases can be used to modify the structure and composition of oils and fats (67). Lipases are available that will catalyze reactions at all three positions on the glycerides, or on the 1,3 positions. Enzymatic interesterification permits greater control and range of specificities than existing technologies. The ability to produce different triglyceride compositions is increasing rapidly.

Beta-glucans are said to have the ability to lower cholesterol levels. Elucidation of the mechanism by which the beta-glucans function would pave the way for application of biotechnology for glucan production.

NEW PRODUCTS: NEW FATS, LOW FATS, AND FAT SUBSTITUTES (NO FATS)

One of the long-term objectives of the Department of Health and Human Services by the year 2000 is to have at least 2000 brand items of processed food products on the market that are reduced in fat, saturated fat, and cholesterol (68).

A high intake of oils and fats can cause weight gain and obesity, but the real importance of fat in our diets lies in the broader, important dimensions of health, disease, and managed nutrition. Hence, a considerable product development effort has been conducted over the past few years related to fat substitutes and new and low-fat products.

This important effort is encountering two very basic aspects. They are as follows: (1) In most if not all instances, and certainly with all fat substitutes, they will not match conventional fat performance, behavior, and acceptance by the consumer; (2) consumers continue to say that there are more important criteria in their purchasing of foods than health and safety, with taste or flavor being the key criterion.

Specific food functions attributable to oils and fats include flavor/aroma, mouth feel, texture, palatibility, smoothness, freshness, moistness, tenderness, creaminess, crispness, satiety, ingredient solubilization and transport, and heat-transfer properties. Be aware that microbiological stability could also be altered. Table 13.1 (69) elaborates on the point more thoroughly via commodities. The authors recommend screening and direct comparison of the full fat product and its intended replacement. The writer judges that flavor (or taste) is also an attribute with frostings and fillings. Successful reformulation comprises the bulk of the article, and possible con-

Table 13.1. Application of starch-based fat replacers.

Baked goods	Salad dressings	Dairy products	Meats	Frozen desserts	Frostings and fillings	Sauces, gravies, and soups
Flavor	Flavor, saltiness, acidity	Flavor, saltiness, acidity				
Viscosity/body	Viscosity/body, yield stress	Viscosity/body, yield stress				
Richness				Flavor	Viscosity	
Texture/grain	Egg flavor, graininess	Smoothness	Flavor	Viscosity/body	Mouthfeel	
Aeration	Smoothness	Creaminess	Mouthfeel	Mouthfeel	Smoothness	
Shortness	Aeration	Mouthfeel	Juiciness	Creaminess	Melt	
Tenderness	Mouthfeel	Satiety	Firmness	Melt	Aeration	Flavor, saltiness
Leavening	Emulsion	Melt	Satiety	Surface appearance	Spread	Mouthfeel
Lubricity	Spreadability	Emulsion	Handling	Overrun	Emulsion	Satiety
Dough handling	Surface appearance	Aeration	Emulsion	Shrinkage		
Batter stability		Surface appearance	Heat transfer			

Source: Food Technology 46(6) 146 (1992).

sulting is suggested. Selection of any replacement for any end-product use will depend on that end product, its properties, and how it is to be used.

The following quotation (70) spells out the "need."

> It now seems that every food company, ingredient supplier, and food formulator is looking for a colorless, odorless liquid or semisolid that looks, tastes, and functions like an oil or fat. Ideally this magical material must be non-caloric, has no adverse nutritional effects, be microbiologically stable, and, preferably it should cost less than water.

The preceding article on "Fat Replacers" (71) contains tables on new product introductions in 1981 and 1991, and maximum potential markets for fat substitutes.

Relative to basic concern number two, recent consumer surveys indicate that taste (flavor), price, variety, and ease of preparation rank higher than nutritional value in consumer response. Product safety has been rated below taste, which is rated as the most important criterion for repurchase (72). It is expected that this ranking could change in view of the increasing emphasis on nutrition.

Chemically, these fat "replacements" can be categorized into two groups: (1) structural analogues of oils and fats, examples being Olestra (a sucrose polyester—the initial development of this type), alkyl glycoside fatty acid polyester, long-chain diolesters, polysaccharide polyesters, polyvinyl oleate, and trialkoxy tricarboxylate. Caprenin also fits in this category as a medium-chain length triglyceride: (2) Biopolymers (proteins, gums, starches, pectins, cellulosics, and beta-glucans) all having a long history of being components of foods. The most widely known example is Simplesse. It is the initial protein-based fat substitute. A number of both types will be mentioned toward the end of this section.

A 1992 article on food additives predicts a 25% annual growth of "fat replacers" from 1991 to 1996, far outstripping growth of all other food additives. Reference to this article is recommended (73). The article also indicated that U.S. consumers (about 75% of adults) were the leaders in consumption of low-fat, low-caloric products (74). From the bar graph in the article, there is no question that United Kingdom consumers buy more of only low-fat products, with the two countries appearing about the same for only low-calorie products. France and Germany were also in the reporting.

The article following "Fat Replacers" is entitled "Emulsifiers and their role in low-fat and no-fat processed foods." It is also worth consulting. Listed are the following 10 categories of fat replacers:

- Emulsifiers or surface active agents
- Synthetic fat substitutes, for example, sucrose polyesters
- Hydrocolloids; for example gums
- Starch derivatives; for example, maltodextrins
- Hemicellulose
- Beta-glucans; for example, oat bran hydrolysates
- Soluble bulking agents; for example, polydextrose
- Microparticulates—size-restricted protein product
- Composites
- Functional blends

Who makes and what are these new products, whether fats, low-fats, or replacers? Table 13.2 lists most if not all as of this writing.

For worthwhile information on the properties of these materials and their applications in products, consult the cited references.

The following new product types have been receiving the most attention:

- Low-calorie protein-based fat substitutes
- Synthetic fats
- Fats from microalgae
- Medium-chain-length triglycerides
- Other synthetic substitutes
- Gums
- Carbohydrate-based substitutes
- Structured triglycerides

Selected examples from these types are discussed in some of the following paragraphs.

The initial protein-based fat substitute is Simplesse (NutraSweet), a sugar, egg white protein, and milk combination with a simulated fat texture. A recent formula change replaced the sugar with NutraSweet artificial sweetener. The product has been marketed as a frozen dessert named Simple Pleasures, now Simple Pleasures Light. (NutraSweet has recently sold this business). Market success had not met expectations, hence a recent formula change and sales emphasis as an ingredient rather than a finished product. Protein products like Simplesse lack heat stability and, therefore, are not suitable in cooking. Two similar products awaiting marketing are Trailblazer

Table 13.2. Fat "replacers."

Company	Product name	Chemical nature	Finished food product Possible uses
AEP Colloids American Lecithin Company	OMI 5050 Alcolec 140	Carrageenan Phospholipid concentrate	Meats Cakes, breads, cookies, pie crusts and other baked goods, salad dressings, margarines, cocoa butter
American Maize Products Company	Amalean I Amalean II	High anylase starch bases	Salad dressings, icings, cream fillings, dairy products, sauces, baked goods
ARCO Chemical Company/ CPC International Inc.	EPG	Esterified propyoxylated glycerol	
Ault Foods LTD	Dairylite	Milk	Frozen desserts, dairy products
Avebe America Inc.	Paselli SA2	Potato-based maltodextrin	Frozen desserts, baked foods, dips, salad dressings, frostings, sauces, gravies
Best Foods, Division of CPC	TATCA	Trialkoxy tricarballate	Not being pursued at this time (late 1992)
Canadian Harvest USA	Snowite Oat Fiber		Baked goods, reduced-calorie breads
Carrageenan Marketing Corp.	Carra Fat	Carrageenan based	Processed meat

Table 13.2. Continued　Fat "replacers."

Company	Product name	Chemical nature	Finished food product Possible uses
Central Soya Company, Inc.	Centrolex line	Deoiled lecithin	Particularly baked goods with soy proteins in meat and dairy
Commercial Creamery Company	Super Creme C Super Creme CSD	Milk protein plus stabilizer	Frozen desserts, baked goods, meat, salad dressings, soups, sauces, yogurt
Con-Agra	Trim Choice (an Oatrim)	Oats, maltodextrin, and glucans	Spreads, salad dressings, frozen desserts, baked goods, cereals, meats, confections
	Leanesse	Modified oat flour	
CPC		Trialkoxy tricarboxylate	Not being pursued at this time (late 1992)
Continental Colloids		Colloid-based stabilizers	Frozen desserts, yogurt, whipped toppings, dairy products
Cumberland Packing Corp.	Dried Cream Extract	Enzyme-modified cream with maltodextrin	Spreads, salad dressings, frozen desserts, baked goods, confections, milk beverages, creamers, gravies, soups, creme fillings and toppings

Table 13.2. Continued   Fat "replacers."

Company	Product name	Chemical nature	Finished food product Possible uses
Delta Fibre Foods	Fibrex	Fiber from sugar beets	Baked goods, meats, soups, snacks, cereals, gravies, sauces, dressings, frozen desserts, confectionery products
Dow Chemical	Methocel	Methyl cellulose gums	Bakery products, fried foods, salad dressings
Excelpro Inc. Eastman Chemical Company	K-pro Myvacet 9-45	Distilled acetylated monoglycerides	Cakes, muffins, cookies
	LITA	Zein protein base	Joint with Pfizer
FMC	Avicel	Carbohydrate-based cellulose	Cheese, cake mixes, frostings, frozen desserts, condiments, spreads
	Nutricol	Konjac flour	Pound cake, other bakery items
	Novogel	Cellulose	Salad dressings, condiments, spreads, gravies, sauces
		Carrageenan	Meats, cheese
Frito Lay	DDM	Dialkyl dihexadecy-malonate	Work continuing (late 1992)

Table 13.2. Continued   Fat "replacers."

Company	Product name	Chemical nature	Finished food product Possible uses
Grain Processing	Maltrin	Cornstarch	Frozen dairy desserts, baked goods, beverages, salad dressings, spreads, meats, cheese, peanut butter, dairy, yogurt
Hercules	Slendid	Pectin	Very low levels can provide up to 100% replacement in a wide variety of cold and heated foods (few calories)
ITC Gums Inc.	Colloid No Fat 102	Gums and starch	Jams, jellies, bakery products, frozen desserts, frostings, salad dressings, meats
Kelco	Kel-Lite	Alginates, xanthan gum, gillangum	
Kraft General Foods	Trailblazer	Modified protein texturizer	
Karlshamms USA		MCT = medium chain triglycerides	

Table 13.2. Continued   Fat "replacers."

Company	Product name	Chemical nature	Finished food product Possible uses
Leiner USA		Gelatin	Spreads, margarine, cookies, muffins
Lifewise Ingredients Inc.	Simply Rich	Carbohydrate bases	Yogurts, salad dressings, ice cream and frozen desserts, cheeses, dips
	Ambiance		Meat and bakery products
Lucas Meyer Inc.	MC-Thin HL66	Modified soy lecithin	Salad dressings, mayonnaise, margarine
	Amisol EC1009	Lecithin, fatty acids, carbohydrate	Baked goods, buns, sweet doughs
	Lecimulthin 100	Powdered lecithin	Baked goods, ice cream
Meer Corp.	Merecol LIP-R		
Mid-America Food Sales Ltd.	Accugel 200	Carbohydrate and protein	Spreads, baked goods, meats, confections, snack foods
National Starch & Chemical Company	N-Lite (6 types)	Maltodextrins and modified food starches	Dairy, baked goods, frostings and fillings, meat sauces, salad dressings

Table 13.2.Continued   Fat "replacers."

Company	Product name	Chemical nature	Finished food product Possible uses
	N-Oil and Instant N-Oil	Tapioca dextrins and maltodextrin	Frozen desserts, puddings, sour cream, salad dressings, table spreads, cheese sauces, meats
	N-Flate	Emulsifiers, guar gum, modified starch, milk solids	Baked products
NutraSweet	Simplesse	Whey protein	Frozen desserts, cheese, baked goods, puddings, salad dressings, dips and spreads, soups, sauces, whipped toppings and frostings
Opta Food Ingredients Inc.	OptaGrade Crystalean	Starch based oat fiber	Fried foods, baked goods, salad dressings
Pfizer	Litesse I & II	Dextrose polymer	Candy bars
	Veri-Lo 100	Water, soybean oil modified starches, agar, monoglycerides and diglycerides, polysorbate 60, K sorbate, phosphoric acid	Extender in wide range of foods

Table 13.2. Continued   Fat "replacers."

Company	Product name	Chemical nature	Finished food product Possible uses
	Veri-Lo 200	Water, anhydrous milk fat, modified starch, agar, monoglycerides and diglycerides, polysorbate 80, K sorbate, phosphoric acid	Extender in wide range of foods
P B Gelatins (Division of Tessenderlochemie)	Slimgel	Gelatin, gums	
Procter & Gamble	Olestra	Sucrose polyester	Snacks
	Caprenin	Caprylic, capric, behenic glycerides	Candy bars
Quaker Oats Company/ Rhone Poulenc	Oatrim (a USDA development)		See Con-Agra
Reach Associates Inc.	Not so Fat C Not so Fat PC	Starches Protein and starches	Salad dressings, ice cream and frozen desserts, yogurt, baked goods, mayonnaise
Shemberg USA		Carrageenan	Meats, frozen desserts, dairy, beverages, pudding and flan mixes, water dessert gels, jellies

Table 13.2. Continued   Fat "replacers."

Company	Product name	Chemical nature	Finished food product Possible uses
A E Staley	Stellar	Cornstarch	Margarines and table spreads, salad dressings, baked goods, cheese, sauces, frostings, fillings, frozen desserts, meats
	Sta-Slim 143	Modified potato starch	Cheesecakes, soups, cream cheese, salad dressings
Tephan	Neobee M-5	MCT	
Tiense Suikerraffinader Services	Raftiline	Inulin (soluble fiber)	Up to 100% fat replacement in a wide range
Tipiah Inc.	Tapicaline	Cassava starch	No-fat and low-fat baked goods, dairy products, processed meats and sauces
Unilever		Sucrose polyester	
Unnamed European Manufacturer	Cerestar	Potato-based maltodextrin	Salad dressings, margarine, ice cream fillings, processed meats
Zumbro Inc.	Rice Complete	Maltodextrin plus protein	Baked goods, salad dressings, mayonnaise, frozen dairy desserts

(Kraft General Foods) and LITA (Enzytech Food Ingredients). The latter is said to have better heat stability and can be used in medium-heat applications.

Sucrose polyesters, which are a category of synthetic fat, are heat stable, not absorbed by the body, hence furnish no calories, and are made from sucrose and 6 to 8 long-chain fatty acids. Initially, the FDA was petitioned for approval of the product, Olestra (Procter & Gamble), for use in shortenings and oils at levels up to 35% in those used in the home and 75% used in foodservice and in snack chips as a partial substitute for conventional fat. That company has requested deferment of that consideration and replaced it by use only in savory snacks such as flavored and unflavored chips and crisps made from potatoes or corn meal, and extruded snacks. Approval awaits congressional action. Extension until the end of 1997 is expected.

Another synthetic fat type which was studied over 40 years ago is acetin fats (Procter & Gamble). Acetic acid was introduced into the triglyceride molecule via interesterification. The purpose was to increase oxidative stability and in shortenings to provide wide plastic ranges. No commercial products have resulted. An offshoot discovery of acetin fats was the acetoglycerides, which have received limited worldwide acceptance as a translucent, edible coating of foods. These completely saturated materials can be "diacetins" or monoacyl, diacetyl triglycerides or the monoacyl, monoacetyl diglycerides. Eastman Kodak has called their products Myvacets.

An exciting new source of conventional fat is from microalgae. These species have the abilities to synthesize specific triglycerides of a more homogeneous nature than the usual fats. Fats of up to 50% of eicosapentaenoïc (EPA) plus docosahexaenoic (DHA) acids have been obtained. But contrary to fish oils, the normal sources of these omega-3 acids, the algae fats lack the normal contamination including off-flavors and toxic substances like DDT and dioxins. Large-scale production is needed to permit greater evaluation.

An alternate fat under consideration for broader usage is medium-chain-length triglycerides (MCT) (Stephan; Karlshamns USA, formerly Capital City Products). C_8 and C_{10} acids, derived from coconut oil, comprise over 98% of the MCT. The fatty acids are metabolized as rapidly as carbohydrates. MCT is not new, having been used for 35 years in certain clinical applications such as malabsorption of fats. Costs are presently too high for common consumption, perhaps partly because of the present medical uses. However, their unusual properties—low viscosity, extreme oxidative stability, no color, bland,

and lower than 8.3 cal/g—are stimulating increased study. Food applications that have been suggested include flavor carrier, confections, reduced-calorie foods, and specialty nutrition, an expansion of their historic role.

Procter & Gamble recently developed a reduced-calorie fat for confectionery manufacturers. It is called Caprenin and is intended to replace cocoa butter in soft candy. It has only 5 rather than the usual 9 cal/g. The constituent fatty acids (caprylic, capric, and behenic) come from peanut oil and butterfat. Behenic is only partially absorbed by the body. FDA has affirmed it as GRAS.

Esterified propoxylated glycerol (Arco) is a low-calorie heat stable fat substitute made by combining propylene oxide with a naturally occurring fat. No petition has been filed.

Dialkyl dihexadecyl malonate (Frito Lay) is a thermally stable synthetic fat substitute currently undergoing animal and human safety testing. It is noncaloric and said to be less than 0.1% absorbed.

Gums are not substitutes per se, but their ability to hold water allows them to partially replace fat while retaining the desired finished product characteristics such as softness. A large advantage is that FDA approval is not needed. McDonald's new low-fat hamburger through the use of carrageenan permits the calories from fat to be reduced from 30% to 9%.

It is also worth noting that ConAgra Inc. introduced a couple of years ago a 96% fat-free ground beef, called Healthy Choice Extra Lean Ground Beef. Healthy Choice is 89% beef. The main additive is a modified oat flour fat substitute called LEANesse. National distribution was the end of November 1991 (75).

Compared to protein-based products like Simplesse, certain carbohydrates used as fat substitutes yield products with heat stability. Most attention has been given to maltodextrins (Grain Processing Corporation; National Starch and Chemical Corporation), which are currently used as fat replacements in salad dressings, margarines, puddings, soups, dairy products, and frozen desserts and are finding applications in baked goods. They are generally recognized as safe (GRAS) by FDA. Other carbohydrate sources being applied as fat substitutes include polydextrose (Pfizer), potato starch (Avibe America Inc.), and tapioca dextrin corn starch (Stellar by A. E. Staley). Recent issues (June 1991, June 1992 and June 1993) of *Food Technology* contain a number of special reports under the heading "New Products & Technologies." Two are on maltodextrin (Oatrim-USDA/ConAgra), and cellulose gel (Avicol-FMC Corp.) (76).

A concept that evidently has received little attention is structured fats. These would appear to be tailor-made conventional type materials with the constituent fatty acids based on nutritional needs. The level of consumed fat probably could be lowered appreciably by this approach, but the costs would seem to be prohibitive for common use. Perhaps that is why it has stimulated little interest. No practical application is known.

Brief mention needs to be made of the appreciable effort and considerable success that has been attained in making use of U.S. domestic oils into replacements for foreign-grown coconut and palm kernel oils. These oils run high in the undesirable C_{12}, C_{14}, and C_{16} saturated fatty acids. The mechanism is via appropriate selection of conditions for hydrogenation. Industries that have been involved include bakery, cereal, confectionery, margarine, vegetable-dairy, and snacks (77).

The reader may want to consult recent articles on the use of replacers in high-ratio layer cakes (78) and their nutritional aspects (79).

"DESIGNER" FOODS

The term "designer" had its origin in the United States in 1989 as the National Cancer Institute launched a 20 million dollar 5-year program to study the anticarcinogenic properties of components of various foods. The food components to be identified are not nutrients, and fats are not greatly involved but the effort is of sufficient nutritional and health potential impact that it bears mentioning. The reader is referred to a series of articles (80) and one appearing in *Food Technology* (81).

The approach is preventive rather than curative; that is, to avoid diseases rather than curing them. The program involves the search for plant materials, phyto chemicals—biologically active compounds—that when added to foods will prevent cancer development.

At least 14 classes of phytochemicals are known or believed to possess cancer-preventive properties. They are carotenoids, coumarins, flavonoids, glucarates, indoles, isothiocyanates, lignens, monoterpenes, phenolic acids, phthalides, phytates, polyacetylenes, sulfides, and tri-terpenes. Sources include flaxseed, soybeans, garlic, green tea, cereal grains, broccoli, cauliflower, brussels sprouts, parsley, celery, carrots, citrus, cabbage, rosemary, and licorice root. Investigation is now concentrating on vegetables and fruits.

International interest and handling vary. Oversight is by the Council for Responsible Nutrition—an international trade organization representing nutrient supplement manufacturers, suppliers, and wholesalers. The Japanese are said to be several years ahead.

Proponents of designer foods believe the first step is to educate the public to eat foods naturally rich in these preventive compounds followed by the creation of safe, marketable products containing the added compounds. Perhaps even use of the compounds in pure form might be reasonable.

Would such materials be foods or drugs? Might they interact? What about possible overuse? What will be the costs? The safety? These and many other questions need to be answered.

The last of the *Inform* articles (80, p. 371) describes "designer" eggs now on limited market in the United States and Canada. These do deal with "oils and fats." They have altered lipid composition—lower saturated, lower total fat, and more omega-3 acids.

NON-NUTRIENT CONSIDERATIONS

Which of the three main classes of food components—carbohydrates, fats, proteins—gives the consumers more pleasure or satisfaction will be left up to the individual consumer. Actually, it is probably a combination of all three. In the judgment of the writer it will either be carbohydrates (sweetness) or fats (satiety and/or other factors such as texture).

The thrust of this section is to present the other factors or product characteristics in which the component fat plays an important role in finished food products, and, if lower in level and/or replaced, quality may be lessened.

Following are definitions of some of the more important ones.

Satiety

Satiety can be defined as the quality or state of being "well-fed, being full." These feelings or sensations more often than not correlate with fat-containing foods.

Appearance

Fats may impart a sheen to some products, a dry appearance to others, and a greasy appearance to some fried foods. Generally, the food processor wants to avoid a greasy appearance, but slightly greasy is preferred with french fries, possibly because that is the usual appearance.

Flavor

Fats will release and enhance the flavors of other ingredients. Fats are the nominal vehicle or carrier of flavors in foods; that is, most flavors are fat soluble. As the formulators decrease the fat level, more bulk via use of complex carbohydrates may be required to carry the flavor.

Generally, sharp-melting fats with lower viscosity release flavors rapidly, whereas higher-melting materials retard flavor release.

Texture

The texture/structure of many food products depends on the interaction of fat solids with other product constituents. Often the structure varies with temperature. Product structure expresses itself in mouth feel, which is critical to a fatty product's acceptance. Therefore, the melting characteristics of the fat ingredients are important in making the final product chewy, soft, smooth, or brittle.

Lubricity

The presence of liquid oil in the formulation of some foods lessens the abrasive effect of other ingredients during mixing. This results in smooth compositions and increased ease of mixing with less energy and time required. Lubricity also has an effect on mouth feel and the ease with which a product can be spread.

Additional organoleptic type qualities which can be associated with fatty foods listed in the literature include: creaminess, crispness, crunchiness, freshness, greasiness, moistness, smoothness, tenderness, thickness, and viscousness. The section on "New Products" in this chapter also speaks to such qualities.

Aeration

Aeration, the addition of air, is promoted by the fat crystals in a shortening. Properly processed plastic shortenings can incorporate large amounts of air during the creaming process. Air entrained uniformly throughout a product, such as a dough, allows for even distribution of leavening gases and the water vapor released during baking. This results in increased volume, lower specific gravity, and desirable texture.

Shelf Life

The shelf-life of products containing fats and oils is related directly to the oxidative stability of those ingredients. In turn, the stability of the fat is related to its fatty acid composition. Saturated fats are the most stable, and unsaturated fats the least stable. Generally, stability decreases as the number of double bonds increases or as the unsaturation increases. The resulting oxidation or rancidity is, in the United States, a very undesirable flavor.

Hydrolysis will occur in most foods in storage. Should the component fats contain even small amounts of lauric acid a soapy flavor will result.

Some additional important considerations are as follows:

- Regulatory approval is required—from the FDA for foods and from the USDA for meat or poultry. This can significantly lengthen the time to market and increase costs. Olestra is an example.

- What will be the demand? It is expected to grow as longevity increases (lesser need for calories) and with increasing emphasis on nutrition.

- An FDA objective via the labeling changes is to encourage product innovation.

- What will the prices be? Will those companies with food lines not market their innovations but merely use in their own products?

- What will be the impact of these raw materials on other product lines, oils and fats per se, consumption of grains, vegetables, fruits? Interestingly, despite the increase of artificial sweeteners in foods, annual consumption of sugar and corn syrups rose from 124 to 139 lbs./capita from 1980 to 1990.

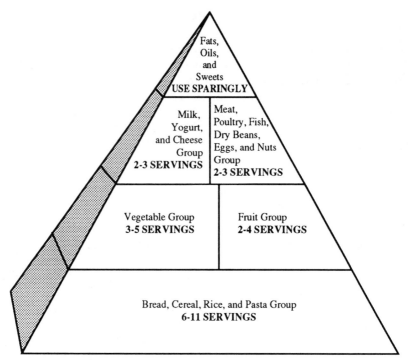

FIGURE 13.1.   Food Guide Pyramid

USDA'S FOOD GUIDE PYRAMID

On April 23, 1992, the USDA released its Food Guide Pyramid. (Fig. 13.1). This new graphic nutritional aid resulted from a collaborative effort of nutritionists, graphic designers, and communication researchers. It replaces the circular graphic aids which stressed four food groups; namely, milk, meat (included eggs, dry peas, seeds, and nuts), fruits and vegetables, and cereal grains.

It is intended for use as an educational tool and by consumers seeking balanced diets. It represents a significant departure on how nutrition is to be taught.

The pyramid reflects the recommended daily consumption by the space allotted to each group. As the graphic depicts, bread, cereal, and so on, are at the bottom (one group—daily servings 6–11), the next largest band is for the fruits and vegetables (two groups—5–9 servings), next to the top are the meat and milk groups (4–6 servings). At the very top are the fats, oils and sweets as one group, to be used sparingly.

As the Secretary of Agriculture (E. R. Madigan) stated at the press conference releasing the pyramid, "All foods are important to a balanced diet" (82). If the pyramid recommends fewer servings of one type or another, that does not mean one food is less important than another. All foods provide important nutrients. The trick is making lower-fat choices.

A 32-page pocket-sized booklet (USDA's Food Guide Pyramid) has been written to accompany the Food Guide Pyramid Graphic Aid. The booklet contains sections on the six food groups, serving sizes, fats, cholesterol, sugars, salt, sodium, and daily intake calorie suggestions. For example, a 2200-calorie diet is considered about right for most children, teenage girls, active women, and many sedentary men. The Pyramid and booklet are available from the USDA Human Nutritional Information Service, 6505 Belcrest Road, Hyattsville, MD 20782.

The *Food Technology* article (82) contains information on the reactions to the Pyramid from various food groups and organizations interested in foods. They were contacted by the Institute of Food Technologists for input. Fifteen such "organizations" returned comments. Twelve were in favor, endorsed, supported. The other three did not indicate they were against; one was uncertain–"doesn't really change the nutritional advice." More extensive quotes from the other two are as follows: (1) From the Institute of Shortening and Edible Oils–"USDA is to be lauded for continuing its nutrition education efforts to teach consumers how to select improved diets. It is hoped the Food Guide Pyramid will indeed play an important role in this process.", and, "While the Food Guide Pyramid takes a major step toward providing consumers with simple, straightforward nutritional information, it may be inherently flawed in that it inadvertently promotes this 'good food/bad food' concept." (2) From the American Meat Institute (AMI), "The revised graphic contains 33 changes from the original 'Eating Right Pyramid' and has been well received by nutrition educators and industry alike. It is important to recognize, however, that no single graphic, no matter how extensively tested, can adequately portray all of the nutrition information contained in the 'Dietary Guidelines for Americans.' " Accordingly, AMI has urged the USDA to develop appropriate nutrition education materials that promote the fundamental concepts of variety, moderation, and proportionality consistent with the guidelines. "We believe the Food Guide Pyramid must be viewed as a single component of a much more comprehensive nutrition education initiative." (The Pyramid was unofficially released in 1991, and the subsequent additional collaborative study cost almost one million dollars).

The responding organizations were American Dietetic Association, American Heart Association, American Meat Institute, Center for Science in the Public Interest, Egg Nutrition Center, IFT—Nutritional Division and a Regional Communicator, Institute of Shortenings and Edible Oils, Inc., National Dairy Council, National Pasta Association, Produce Marketing Association, Rice Millers Association, Tuft's University Diet and Nutrition Letter, and the Wellness Center of Minnesota.

One final quote from the Center for Science in the Public Interest bears sharing. It is "The deficiency is that the top of the pyramid is sweets and oils. That doesn't teach people that most of the fat in an average person's diet comes from meat and dairy." (*Washington Post*). More recently, seen in another press release (83), the Center suggested a color-coded, fold-together Healthy Eating Pyramid which classifies foods in three groups: anytime—low in fat and saturated fat; seldom—very high in fat, saturated fat, or both; and, sometimes—all other foods.

TOTAL DIET RECOMMENDATIONS

An earlier section dealt with level of fat in the diet and spoke to a recommended maximum of 30% of total calories. The following summary is drawn from a comprehensive National Research Council report in the book entitled *Diet and Health: Implications for Reducing Disease Risk* (see Acknowledgments).

Cut back on calories; eliminate excess body fat. Seek the needed calories from 30% (or less) fat, 55% carbohydrates, and 15% protein, with the fat reduction from a current level of about 37% having the highest priority.

Reduce saturated fatty acid intake to less than 10% of calories, and the intake of cholesterol to less than 300 mg daily. The intake of fat and cholesterol can be reduced by substituting fish, poultry without skin, lean meats, and low-fat or nonfat dairy products for fatty meats and whole-milk dairy products; by choosing more vegetables, fruits, cereals, and legumes; and by limiting oils, fats, egg yolks, and fried and other fatty foods.

Every day eat five or more servings of a combination of vegetables and fruits, especially green and yellow vegetables and citrus fruits. Also, increase intake of starches and other complex carbohydrates by eating six or more daily servings of a combination of breads, cereals, and legumes (to provide the desired dietary fiber.) A serving is equal to a half cup for most fresh or cooked vegetables, fruits, dry or cooked cereals and legumes, one medium piece of fresh fruit, one slice of bread, or one roll or muffin.

Maintain protein intake at moderate levels. There are no known benefits and possibly some risks in consuming diets with high animal protein. A majority of Americans consume a considerable excess of the daily requirements for protein. This is costly and unneeded. The daily requirement is 0.8g/kg of desirable body weight. For a 160-lb individual, it is 58 g or 2 oz. It is recommended that the intake be less than twice this level or less than 4 oz/day. Hopefully, this will convince readers that we unnecessarily indulge ourselves in protein.

Consume all food groups. Risk of mortality is lower. This comes from the study of the data from the First National Health and Nutrition Examination Survey.

Do not consume alcohol. For those who do drink alcoholic beverages, the recommended consumption is the equivalent of less than 1 oz of pure alcohol in a single day. This is equivalent to two cans of beer, two small glasses of wine, or two average cocktails. Pregnant women should avoid alcoholic beverages.

Limit total daily salt (sodium chloride) intake to 6 g or less. The American Heart Association recommends a maximum of 3 g of sodium or 7.6 g of salt. Limit use in cooking and avoid adding it to the food at the table. Salty, highly processed salty, salt-preserved, and salt-pickled foods should be consumed sparingly. The food industry needs to control its levels of added salt and other sodium-containing additives more stringently.

Maintain adequate calcium intake. The use of calcium supplements is not justified at this time unless the individual voluntarily or for medical reasons is restricting the intake of calcium-containing foods. Consume low-fat or nonfat dairy products and dark green vegetables.

Avoid taking dietary supplements in excess of the RDA in any one day. The National Research Council Committee on Dietary Allowances stated they found no evidence of beneficial or harmful effects from this general practice in the United States. It emphasized that the long-term risks and benefits of taking supplements have not been adequately studied.

Maintain an optimal intake of fluoride, particularly during the years of primary and secondary tooth formation and growth. This significantly reduces dental caries and no adverse effects on health have been noted.

The USDA Pyramid discussed in an earlier section should also be consulted.

Two individual researcher contributions that may be of passing interest are as follows:

• The individuals dietary intake should contain about 28% fruits and vegetables, 32% cereals and grain products, 19% milk products, and 21% animal and legume foods.

• Three diet change recommendations (84) for reducing the overall risk of a heart attack by up to 46% are shown in the following chart.

Diet Change	To Reduce Blood Cholesterol by	To Reduce Risk of Heart Attack by
Lower saturated fat (10% of calories) and cholesterol (less than 300 mg/day)	10%	20%
Attain ideal weight	10%	20%
Add fiber (4–5 servings/ day of fruits, vegetables or legumes)	3%	6%
		Total 46%

A recent California study of 10,000 Mormon priests and their wives found cancer rates that were 34–50% lower and rates of heart disease 14–34% lower than those of the general population. Life expectancy was 85 for men (compared to the national average of 74) and 86 for women (compared to 80 nationally). Cited among the Mormon's life-style habits were a well-balanced diet, abstinence from tobacco, alcohol, drugs, and caffeine, and a strong family life (less stress) (85).

In the past couple of decades, Americans have been cutting down on red meats, eating more poultry, fish, grains, and cereal products, cutting back on whole milk, and increasing the consumption of fresh fruit and vegetables. These trends need to continue.

NUTRITIONAL LABELING

Background

The United States Nutritional Labeling and Education Act of 1990 (NLEA) required the United States Department of Health and Human Services (HHS) to propose new nutrition labeling regulations by November 8, 1991 and issue final regulations by November 8, 1992, using comments gained from the 1991 release. The act requires quantitative declarations of total fat, saturated fat, and calories from fat.

Regulations were released on time (1991), but they lacked the revisions recommended by the FDA and most of the food industry—more was wanted on health concerns, fat, types of fat, and total diet. More than 40,000 comments were received from industry, trade associations, professional societies, academia, and the public. Many were form letters.

A disagreement arose between FDA and USDA's Food Safety and Inspection Service (FSIS) primarily over the labeling format, including the handling of calories and fat. There was even some talk that finalization should be put off until the new administration could handle it (86). The disagreement was not resolved on time so the regulations became final on November 8, 1992, but nobody was happy. President Bush and his staff entered the picture and the agreed on final regulations were announced on December 2, 1992.

Should the reader want details on the specifics of concern, consult Refs. 87 and 88. The latter of those two references and *The Referee* (89) provide details on the regulations. An FDA Consumer Special Report entitled "Focus on Food Labeling" provides a wealth of details on the entire effort (90).

A just received issue of *Food Technology*, in late July, 1994, contains a regulatory update on nutritional labeling. (91)

New Regulations

On January 6, 1993, the long-awaited regulations were published in the *Federal Register*. They take up 961 printed pages—900 for FDA's final regulations and 3 proposals, and 61 pages for the FSIS's one final regulation and one proposal.

The new regulations specify what information must be on the nutrition panel, define terms that can be used to describe nutrient con-

Table 13.3. Mandatory and Optional Dietary Components in the order in which they must appear on the Nutrition Facts panel. Mandatory components are shown in boldface.

Total calories
Calories from fat
Calories from saturated fat
Total fat
Saturated fat
Polyunsaturated fat
Monounsaturated fat
Cholesterol
Sodium
Potassium
Total carbohydrate
Dietary fiber
Soluble fiber
Insoluble fiber
Sugars
Sugar alcohol
Other carbohydrate[a]
Protein
Vitamin A
Vitamin C
Calcium
Iron
Other essential vitamins and minerals

[a]The difference between total carbohydrate and the sum of dietary fiber, sugars, and sugar alcohol (if declared).
Source: *Food Technology 47* (2), 82 (1993).

tent, allow certain health claims about the relationship between nutrients or foods and diseases, standardize serving sizes, require full ingredient listing, and require declaration of total percentage of juice in juice drinks. Table 13.3 lists the mandatory and optional dietary components in the order they must appear on the Nutrition Facts label. Mandatory components are in boldface.

Effective by May 1994

The NLEA specifies that food labels will have to meet the new requirements by May 1993 but gives FDA the authority to postpone by up to a year if the final regulations cause undue hardship. The FDA decided on the compliance date of May 8, 1994, except for certain regulations pertaining to health claims and ingredient labeling. Congress, in late May, 1994, granted a delay until August 18, 1994, with extensions only granted upon request. The extension was requested by food packagers to assist their using up outdated wrappers. The effective date for labeling of meat and poultry products was July 6, 1994.

The FDA has estimated that each of more than 257,000 different food products and package sizes will need to be relabeled. FSIS estimates that more than 90% of all processed meat and poultry products will require nutrition labeling.

Estimated Costs

The FDA has estimated that the new food labeling will cost FDA-regulated food processors 1.4–2.3 billion dollars and the government 163 million dollars over the next 20 years. Benefits to public health are expected to well exceed such costs.

Exemptions

Most foods will require nutrition labeling. The following are exempt but may carry nutrition information as long as it complies with the new regulations: food sold by small businesses; food sold in restaurants and other establishments for immediate consumption; ready-to-eat foods prepared on site for later consumption; foods that contain insignificant amounts of nutrients, such as coffee, tea, and spices; dietary supplements, except in food form; infant formula; medical foods; custom-processed fish and game; foods shipped in bulk but not sold to consumers in that form; donated foods; meat and poultry products in small packages weighing less than $\frac{1}{2}$ oz; and meat and poultry products produced or packaged at retail, such as sliced bologna. If these foods make a nutritional or health claim, they must bear nutritional labeling.

Be aware that small businesses, restaurants, and dietary supplements have been studied further hence changes have been made and will be made in at least these three areas. As Dr. F.E. Scarbrough, Director of the Office of Labeling of the FDA has said, "Food labeling is a subject we'll continue to revisit for many years to come" (90). Nutrition labeling is a very fluid area. The reader needs to be aware that whereas every effort was made to make the presented information accurate and up-to-date prior to publication, it will not be so at the time of reading (see reference 91 for an update after this chapter was written).

Fresh fruits and vegetables and raw fish, as of this writing, do not need to be labeled. Although the law mandates nutritional labeling for almost all processed foods, it allows voluntary points-of-purchase nutrition information for raw fruits, vegetables, and fish—as long as a sufficient number of retailers participate. In a voluntary FDA program, nutrition information on the 20 most frequently used foods in each of these programs is to be on display at point of purchase in food stores. The FDA will survey this, and if compliance is less than 90%, then mandatory labeling will be required.

The FSIS exempts raw single-ingredient meat and poultry products. An example is chicken breasts.

Advertising must follow the rules or guidelines as required in the labeling.

The New Nutrition Panel

Figure 13.2 shows the new panel. No components other than those listed in the table can be on the label. Any listed components must have nutritional information on them.

Simplified formats can be used in certain instances like insignificant levels of mandatory nutrients. Information on total calories, total fat, total carbohydrate, protein, and sodium must be present.

Small packages of less than 12 in.^{2} available for labeling may omit labeling, but an address or phone number must be provided so consumers can obtain the information.

Vitamins/Minerals

Heart disease, cancer, diabetes, and many other potential ailments due to poor nutrition and overeating have supplanted the concerns over vitamins and minerals.

Vitamins A and C continue to be required because it is believed they minimize cancer risk. Calcium remains in view of its importance in avoiding osteoporosis. Similarly, iron is still present as it helps prevent anemia.

The other essential vitamins and minerals are optional unless claims are made for either or the food is fortified or enriched with either.

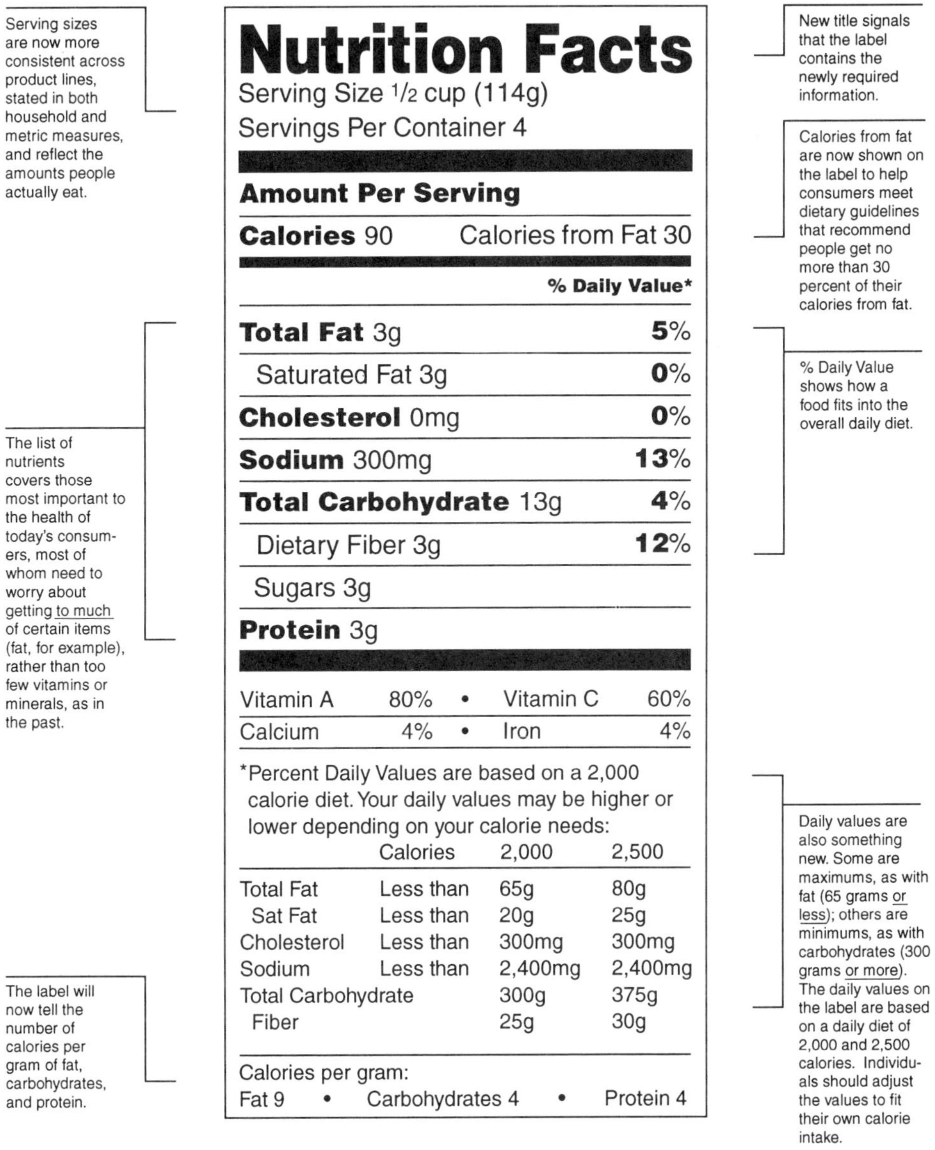

Serving sizes are now more consistent across product lines, stated in both household and metric measures, and reflect the amounts people actually eat.

The list of nutrients covers those most important to the health of today's consumers, most of whom need to worry about getting to much of certain items (fat, for example), rather than too few vitamins or minerals, as in the past.

The label will now tell the number of calories per gram of fat, carbohydrates, and protein.

Nutrition Facts

Serving Size 1/2 cup (114g)
Servings Per Container 4

Amount Per Serving

Calories 90   Calories from Fat 30

	% Daily Value*
Total Fat 3g	**5%**
Saturated Fat 3g	**0%**
Cholesterol 0mg	**0%**
Sodium 300mg	**13%**
Total Carbohydrate 13g	**4%**
Dietary Fiber 3g	**12%**
Sugars 3g	
Protein 3g	

Vitamin A	80%	•	Vitamin C	60%
Calcium	4%	•	Iron	4%

*Percent Daily Values are based on a 2,000 calorie diet. Your daily values may be higher or lower depending on your calorie needs:

		Calories	2,000	2,500
Total Fat	Less than	65g	80g
Sat Fat	Less than	20g	25g
Cholesterol	Less than	300mg	300mg
Sodium	Less than	2,400mg	2,400mg
Total Carbohydrate		300g	375g
Fiber		25g	30g

Calories per gram:
Fat 9   •   Carbohydrates 4   •   Protein 4

New title signals that the label contains the newly required information.

Calories from fat are now shown on the label to help consumers meet dietary guidelines that recommend people get no more than 30 percent of their calories from fat.

% Daily Value shows how a food fits into the overall daily diet.

Daily values are also something new. Some are maximums, as with fat (65 grams or less); others are minimums, as with carbohydrates (300 grams or more). The daily values on the label are based on a daily diet of 2,000 and 2,500 calories. Individuals should adjust the values to fit their own calorie intake.

FIGURE 13.2.   Nutrition facts [from Food Technol. 47(2), 83 (1993)].

An Example of the New Nutrition Panel, *which will appear on almost all packaged foods (compared to about 60% of products at present). The label shown is only a sample. Exact specifications are in the final regulations. Figure courtesy of the Food and Drug Administration.*

Final regulations for vitamins and minerals were published in January, 1994. The final rule basically treats dietary supplements as foods. However for vitamins and minerals, which have RDIs and DRVs— see next topic, the new rule requires labels that are similar yet different from those required for conventional foods. All other types, which don't have RDI and DRVs, of dietary supplements must use the same labeling format as conventional foods. Full compliance is scheduled for July 1, 1995.

RDA

The RDAs (Recommended Daily Allotments) have been renamed RDIs or Reference Daily Intakes. The FDA made this change to eliminate confusion with the National Academy of Sciences RDAs which stand for recommended dietary allowances. The latter are updated periodically to reflect current scientific knowledge.

The recommended daily intakes (RDIs) are a set of dietary references based on the RDAs for essential vitamins and minerals, and, in selected groups, protein. The name, RDI, replaces the term U.S. RDA. A few RDIs are: vitamin A 5,000 International Units; vitamin C 60 mg; calcium 1.0 g; and, iron 18 mg.

Daily Reference Values (DRVs) are a set of dietary references that apply to fat, saturated fat, cholesterol, carbohydrates, protein, fiber, sodium, and potassium. These values are, respectively, 65 g, 20 g, 300 mg, 300 g, 50 g, 25 g, 2,400 mg, and 3,500 mg, for a 2,000 calorie per day intake for adults and children over 4, only.

Daily Values (DVs) are a new dietary reference term made from DRVs and RDIs. It is to help consumers use food labeling information to plan an overall healthy diet. Only the DVs will appear on labels. It will help consumers in that (1) it serves as a basis for declaring the percent of the Daily Value for each nutrient that a serving of the food provides and (2) it provides a basis for thresholds that define descriptive words for nutrient content, called descriptors, such as high fiber and low fat. As an example, the descriptor "high fiber" can be used if a serving of food provides 20% or more of the Daily Value of fiber, that is, 5 g or more. The Daily Values are not recommended intakes but reference points.

The FDA has been considering revising the values.

Descriptors

The FDA and FSIS have defined descriptive terms, such as free, reduced, light, and more, to minimize consumer confusion. A listing of these is given in Table 13.4. The information will not be current at the time of reading. As examples, reference 90, p. 29, lists 11 claims which are called core terms, 10 of which are in the table, and, the definition of health was undergoing revision.

Manufacturers may petition the agencies for the use of additional descriptors.

Health Claims

Claims for eight relationships between a nutrient or food and the risk of a disease or health-related condition will be allowed for the first time. These are the following:

- Calcium and osteoporosis
- Fat and cancer
- Saturated fat and cholesterol and coronary heart disease
- Fruits, vegetables, and grain products and cancer
- Fruits, vegetables, and grains and coronary heart disease
- Fruits and vegetables and cancer
- Sodium and hypertension
- Folic acid (part of vitamin B complex) and neural tube birth effects

No other health claims are allowed.

Claims may be made through third-party references, but such claims must meet certain requirements.

Analytical Considerations (See Chapter 14)

The presence of the nutrient descriptors places burdens on the analytical chemist and the food industry. Revised and new methods will be needed.

Particularly as the world moves to a global economy, agreement on not only methods but definitions will be needed. Fat, via the regulations, was defined as the sum of the fatty acids as expressed as triglycerides, including free fatty acids, fatty acids from monoglycerides and diglycerides as well as triglycerides and any other source.

Table 13.4.   Nutrient content descriptors that may be used on food labels.

Descriptor[a]	Definition[b]
Free	A serving contains no or a physiologically inconsequential amount; <5 calories; <5 mg of sodium; <0.5 g of fat; <0.5 g of saturated fat; <2 mg of cholesterol; or <0.5 g of sugar
Low	A serving (and 50 g of food if the serving size is small) contains no more than 40 calories; 140 mg of sodium; 3 g of fat; 1 g of saturated fat and 15% of calories from saturated fat; or 20 mg of cholesterol; not defined for sugar; for "very low sodium," no more than 35 mg of sodium
Lean	A serving (and 100 g) of meat, poultry, seafood, and game meats contains <10 g of fat, <4 g of saturated fat, and <95 mg of cholesterol
Extra lean	A serving (and 100 g) of meat, poultry, seafood, and game meats contains <5 g of fat, <2 g of saturated fat, and <95 mg of cholesterol
High	A serving contains 20% or more of the Daily Value (DV) for a particular nutrient
Good source	A serving contains 10–19% of the DV for the nutrient
Reduced	A nutritionally altered product contains 25% less of a nutrient or 25% fewer calories than a reference food; cannot be used if the reference food already meets the requirement for a "low" claim
Less	A food contains 25% less of a nutrient or 25% fewer calories than a reference food
Light	1. An altered product contains one-third fewer calories or 50% of the fat in a reference food; if 50% or more of the calories come from fat, the reduction must be 50% of the fat); or 2. The sodium content of a low-calorie low-fat food has been reduced by 50% (the claim "light in sodium" may be used); or 3. The term describes such properties as texture and color, as long as the label explains the intent (e.g., "light brown sugar," "light and fluffy")
More	A serving contains at least 10% of the DV of a nutrient more than a reference food. Also applies to fortified, enriched, and added claims for altered foods
% Fat Free	A product must be low fat or fat free, and the percentage must accurately reflect the amount of fat in 100 g of food. Thus, 2.5 g of fat in 50 g of food results in a "95% fat-free" claim

Table 13.4. Continued   Nutrient content descriptors that may be used on food labels.

Descriptor[a]	Definition[b]
Healthy	A food is low in fat and saturated fat, and a serving contain no more than 480 mg of sodium and no more than 60 mg of cholesterol
Fresh	1. A food is raw, has never been frozen or heated, and contains no preservatives (irradiation at low levels is allowed); or 2. The term accurately describes the product (e.g., "fresh milk" or "freshly baked bread")
Fresh frozen	The food has been quickly frozen while still fresh; blanching is allowed before freezing to prevent nutrient breakdown

[a]See the regulations for acceptable synonyms.
[b]These definitions have been simplified for this table; see the regulations for specific restrictions and additional requirements.
[c]Proposed by FSIS for meat and poultry products.
Source: *Food Technology 47* (2) 85 (1993).

In the United States, saturated fats are defined as láuric through stearic. In Canada and the European Community, it is all "fats" without double bonds.

In the United States, fat free is <0.5 g per serving; in Canada, it is 0.1 g, and in the United Kingdom, 0.15 g. Low fat in the United States at 3 g/serving also differs from Canada and the European Community.

Reference 91 contains a report by the AOAC task force on p. 73.

Fat-Labeling Issues

The AOCS at its 1992 annual meeting in Toronto held a session on "Issues in Food Labeling" (87). Worth mentioning are the following:

- Descriptions of nutrients will hamper new product development and delay move to global economy by limiting trade.
- Needing to be resolved is labeling for unsaturated fat and for trans acids.
- The fatty acid profile of the United States is changing.
- What about chemical versus physiological end points.

- Dietary recommendations differ for the prevention of heart disease versus cancer versus immune dysfunction.

From that same meeting a quote from K. E. Gowrlie, Director of the Consumers Product Branch of Consumer and Corporate Affairs, Canada, is of interest. "The NLEA is a complex and advanced approach to food labeling, nutrition information, and education, but it may also prove to be one of the most trade-limiting, nontariff barriers in global food trade."

For further details aimed primarily at fat see a review article in a question-and-answer format (92).

For information on Nutritional Labeling in Food Service see the April issue of *Food Technology* (93). The actions of the top six food service companies, in sales, are given.

The FDA has proposed amending its regulations by removing the provisions that exempt restaurant menus from the requirements for how nutrient content claims and health claims are to be made. These decisions are considered significant since Americans spend about one-half of their food dollars in restaurants. Both could be finalized by late 1994.

What do consumers look for when they read food labels? A survey in 1993 by the Food Marketing Institute of 1000 shoppers indicated fat content was no. 1, followed closely by ingredients, calories, and other nutrient information (94).

There has been talk by FDA that it will conduct its own survey.

CURRENT/FUTURE RESEARCH

The optimal determinator of the mix of the fatty acids in the diet is a major health need; hence, continuing investigation is required.

Further investigation is needed to confirm and extend recent findings with particular emphases on the identification of foods and dietary components that alter the risk of chronic disease; to elucidate the mechanisms of actions; and to improve the methodology of collecting and assessing the pertinent data.

Additional basic research is needed to increase the understanding of the processes involved in the development of atherosclerosis and the factors influencing them.

More research is needed to elucidate the precise relationships among different classes of fatty acids, lipoproteins, and coronary heart disease mortality.

The significance of high levels of cholesterol in the elderly, including postmenstrual women, needs additional study.

Further exploration is needed of the relationship among genetic factors, dietary fats, and carcinogenesis including individual dietary responses. Methods are needed to determine more accurately who is at risk and who would benefit from intervention.

Research aimed at finding ways to prevent or reverse athero sclerosis is in progress. One of the most promising areas of research is in finding ways to control elevated levels of cholesterol and other fats in the blood.

Three fairly recent findings have been promising. One is the discovery of cell-surface receptors for LDL. These receptors, principally in the liver, bind LDL circulating through the blood-stream, allowing the LDL and its cholesterol to enter the liver and other cells. Research has shown that the number of these receptors on cell surfaces is reduced by genetic or dietary factors and blood levels of LDL increase. This increase can lead to more cholesterol being available for deposit in artery walls. Another research finding has resulted from the Coronary Primary Prevention Trial (CPPT). It showed that lowering a high level of blood cholesterol by means of the drug cholestyramine reduced the risk of death from a heart attack. Additional drugs used to lower cholesterol include lovactatin and probucal. A number of studies on diet/drug interventions to lower cholesterol indicated regression of cardiovascular lesions is possible.

A final very important advance in recent years has been the development of a new class of cholesterol-lowering drugs. These compounds act by blocking the synthesis of cholesterol by the body's cells. These are inhibitors of 3-hydroxy-3-methyl-glutaryl-coenzyme A (HMG-CoA) reductase.

Of course many fundamental questions remain. Medical scientists are continuing to search for answers by studying life at its most basic level—the cell.

Even though much more work needs to be done, scientists have found some answers. For instance, they have found a definite relationship between the amount of cholesterol in the bloodstream and coronary artery disease (blockage of the arteries supplying blood to the heart muscle itself). A large body of scientific evidence shows that a diet high in saturated fats and cholesterol can raise blood cholesterol levels and this may contribute to development of atherosclerosis.

That is why the American Heart Association recommends eating foods low in saturated fats and cholesterol. Such a diet is one way to reduce the risk of a high blood cholesterol level and, thus, the risk of heart attack.

Canola oil has been a true success story. To take rapeseed, the oil of which has shown toxic manifestations in animal studies, and obtain an exceptionally nutritious oil with no adverse properties is, indeed, an accomplishment. Biotechnology promises more of such developments plus seeds that are insect, disease, and herbicide resistant.

Specific projects that are either underway or planned include the following studies:

- The removal of dietary cholesterol from milk fats
- Alpha-linolenic metabolism
- Optimum dietary levels of monounsaturated and the N-3 and N-6 polyunsaturated acids
- Additional studies on the effect of trans acids on blood lipid levels
- Deleterious effects of high levels of the essential fatty acids
- Value of fish oils in inhibiting tumor growth
- Influence of LDL in the development of cholesterol plaque
- Reclaiming used fats for animal feed

Two years ago at a workshop on "Dietary Fatty Acids and Thrombosis" (63) the lead speaker was Dr. W. H. Glinsmann, Associate Director for Clinical Nutrition at the FDA. In addition to dealing with questions relative to dietary fatty acids, many of which have been handled by nutritional labeling, Glinsmann also touched on the development of fat-modified products. He emphasized the need to identify the appropriate target populations as well as populations at potential risk. He used the term "nutriceutical" in referring to products (such as those that might be enriched with omega-3 fatty acids) that could provide both essential nutrients and therapeutic value. When considering products for general food use, Glinsmann noted that issues of safety are paramount. When the product is intended for therapeutic use, a risk–benefit analysis is appropriate.

SUMMARY

The nutritional aspects of fats are exceedingly complex and involve many factors including nondietary ones.

It is quite clear that dietary fat exceeds both protein and carbohydrates in importance to good health insofar as possible adverse effects are concerned. The desire to eat fatty foods is probably mainly a sensory one for flavor, texture, and satiety. Yes, we need fat for caloric purposes and the essential fatty acids, but we need to be aware of and try to avoid the negative factors of cardiovascular diseases, cancer, obesity, and others. As the 1989 National Research Council report on Diet and Health (21) stated:

> There is clear evidence that the total amounts and types of fats and other lipids in the diet influence the risk of atherosclerotic cardiovascular diseases and, to a less well-established extent, certain forms of cancer and possibly obesity. The evidence that the intake of saturated fatty acids and cholesterol are causally related to atherosclerotic cardiovascular diseases is especially strong and convincing.

This remains accurate. Total calories, more than fat intake, appear to increase one's chances of getting cancer.

Dietary changes to lower the plasma levels of total cholesterol and LDL in the United States would be desirable. Two approaches can be used to accomplish this. The first is the public health approach, shifting the distribution of cholesterol levels in the entire population to a lower range via dietary modification, exercise, and weight control. The second is a high-risk-patient-based approach that seeks to identify individuals at high risk and bring them into medical intervention. These two approaches are complimentary and together represent a coordinated strategy to reduce cholesterol levels and coronary risk.

There is good evidence for a continued lowering of total cholesterol in the United States; average value now said to be 205 as compared to 213 in 1978. Too low a level, less than 160, may present health risks. This remains under debate. Low cholesterol levels associated with greater near-term mortality may, instead, be a result of cancer or other conditions.

Coronary heart disease is considered most likely when the blood cholesterol level is elevated, and one or more risk factors such as smoking, high blood pressure, diabetes, and family history at an early age are present.

Although the evidence relating dietary fat to cancer is weaker than that relating dietary fat to cardiovascular diseases, the conclusions for both of these major chronic diseases are entirely congruent. Thus, recommendations to reduce the dietary intake of total fat and of saturated fatty acids will lead to reductions in coronary risk and perhaps cancer risk.

Reductions of total fat should be to 30% or less of total calories and of saturated fatty acids to less than 10%. This is what the American Heart Association recommends. The National Cholesterol Education Program (NCEP) says less than 30%.

The NCEP has also recommended focusing on intakes of saturated fats and cholesterol and that diet, physical activity, and loss of weight should precede possible drug therapy. Dietary guidelines are considered the same for heart disease, cancer, and obesity.

Palmitic acid, because of its abundancy, is the most deterious saturated fatty acid. There is some evidence that myristic acid could be worse on a weight basis.

Studies continue to show that monounsaturated fatty acids (oleic) lower LDL, the "bad" lipoprotein.

Lp(a) lipoprotein has reemerged as a blood protein of interest and concern.

Essential fatty acids appear vital to the proper development of the fetus and children.

Routine measurement of waist : hip ratios of adult women is recommended as a better means of assessing cardiovascular risk factors than body weight.

There is growing interest in antioxidant nutrients and designer foods as preventive health measures.

Ingestion of fruits and vegetables and soybean products are said to reduce risk to certain types of cancer.

The intake of fried foods should be moderated. It follows that the industry probably needs to check the quality of its in-use frying fats more stringently and to control the frying operation to minimize fat levels in the fried products.

Highly oxidized and heat-abused fats could have toxic manifestations. Evidence indicates that off-qualities result first, so ingestion is unlikely. Conventional handling and use do not produce toxic byproducts. However, oxidation of lipids in the body, frequently termed lipid peroxidation, is growing in concern as having a role in chronic disease development.

The effects on blood lipids of trans fatty acids have recently been questioned again. Most of the recent clinical studies on trans acids have shown that they raise total cholesterol, LDL, and Lp(a) and lower HDL. This has led the FDA to say it may consider trans acid labeling in the near future.

Food technologists need to keep in mind that lowering the fat content of foods can have negative influences on product appearance, eating qualities, flavor, and so on.

Sections are included on the USDA Food Guide Pyramid, nutritional labeling, biotechnology, antioxidant nutrients, new "fat" products, and "designer" foods.

KOSHER DIETARY GUIDELINES

The kosher dietary laws, sometimes referred to as the *kashruth*, are observed to varying degrees by members of the Jewish faith. Further, many nonobservant Jews and non-Jews prefer kosher products and kosher meals because they perceive them to be of higher quality. Other groups, such as the Seventh Day Adventists, use kosher regulations and kosher markings of products as guides for their own observance or preference.

This subject is, therefore, of extreme importance to the food technologist who is involved with any food processing plant, where kosher food products are being made or are contemplating being made.

It has been estimated that more than 21,000 kosher-supervised products are now available, that 30% of products in a typical supermarket are kosher, and that "kosher" customers constitute only about 25–30% of the kosher market (95).

A general knowledge of the kosher dietary laws is also important to many who are charged with menu responsibility and/or quantity recipe development in foodservice.

In foodservice applications, there is a growing demand for kosher meals for both religious and quality (including cleanliness and sanitary) reasons. Areas of greatest demand include airline catering, hospitals, nursing homes, office and plant catering, and restaurants and cafeterias in locations with large Jewish populations. These locations include the following metropolitan areas: New York City, Chicago, Los Angeles, and Miami.

Currently, the three major divisions of Judaism are Orthodox, Conservative, and Reform. The Orthodox are the most traditional and they are the most strict in following these dietary regulations. The Conservatives occupy the middle ground, and the Reform group is the most liberal in following kosher dietary regulations. To further complicate the picture, there can be differences of opinion and practices among rabbis, especially between the Orthodox and Conservative rabbinates.

This particular discussion should be considered as merely an introduction to the subject, and it will emphasize the role of fats and oils in these regulations.

There are considered to be three types of names for foods that have received some degree of control or acceptance by the Jewish faith. "Pareve" is the least strict, followed by "kosher," which generally is endorsed by a kosher label or an insignia provided by rabbis or rabbinical councils that provide this service. Foods certified as "kosher for Passover" are produced under the most stringent conditions.

Pareve

Jewish dietary laws state that meals with meat cannot be served with dairy products and vice versa. However, either may be served with vegetables or with other nonmeat or nondairy products that are considered to be pareve. "Pareve," therefore, indicates a neutral food product that may be consumed with either meat or dairy products.

Vegetable oils and fats can generally be considered as pareve. Most meat fats and butterfat could not be considered pareve.

The first step in foodservice operations that have an interest in serving kosher meals is to separate meat and dairy products. The rules of kosher food preparation prohibit dairy foods and meats from being cooked in the same pots and pans or served with the same plates and utensils.

Kosher

Kosher foods are prepared according to Jewish dietary law. In food plants, foods must be processed and handled with utmost care. The origin and prior handling of all ingredients must be known. It is necessary to ensure that all of the equipment used in processing is properly koshered, and there must be no comingling with any prohibited items at any step in processing. Rabbinical supervision is necessary. In some food plants, rabbinical inspection and certification at regularly scheduled intervals are all that is required. In more complex food plant operations, rabbinical inspection is continuous; and a qualified rabbi is on hand at the plant at all times.

Food processing plants must make special efforts to comply with regulations and inspections. Plants producing both kosher and non-kosher foods must keep the two completely separate. The complete procedures for koshering equipment in a plant or foodservice operation are somewhat complex, and one should consult with a rabbi involved in kosher certification to be sure that they are done correctly. One of the most prominent organizations involved in certifications of this type is the Union of Orthodox Jewish Congregations of America, headquartered in New York City. It will conduct the proper supervision of the processing of products and then award a certification of approval that is generally effective for a year. A company that has this certification may use this group's symbol of certification—Ⓤ, or "circle U"—for the products or ingredients certified. Another organization that performs similar duties is the Chicago Rabbinical Council.

Vegetable shortenings and oils that are manufactured under the proper supervision and inspection can qualify for this kosher certification. Meat fats, of course, cannot qualify. In some cases a manufacturer of vegetable shortenings and vegetable salad oils may find that it is more efficient to obtain kosher certification for all products made in a specific plant. In plants where both meat fats and vegetable oils are processed, the two must be completely separated. Foodservice operations must set up a workable arrangement for handling both kosher and nonkosher meals.

Passover Foods

Foods designed for consumption during the week of Passover present even greater problems for the food processor and for foodservice. For instance, they cannot use oil extracted from legumes, such as soybean oil (the most widely used oil in the United States). For Passover, vegetable shortenings and salad oils are generally processed from cottonseed oil.

It should be pointed out that kosher laws did not originate as health laws, but came about more as ethical or moral laws two and a half millenia ago. However, today there is a quality or health connotation, especially among non-Jews who prefer kosher products.

The following are some major concepts concerning kosher laws. They should not be considered to be exact but only as guidelines. Exact rules and regulations must be obtained through the proper rabbinical certification authorities.

Plants

Most plant materials grown in the United States can be koshered. This includes the oils extracted from these plants. However, there are some fine points of law relating to certain agricultural practices that must be clarified.

Meat

The mammals generally considered acceptable for koshering are those that chew their cud and have split hooves. To be included in the Orthodox diet, they must be properly killed and processed.

Milk

As mentioned previously, dairy products must not be consumed with meat products. Dairy products include milk, butter, ice cream, cheese, coffee cream, and so on. Jewish law specifies the appropriate time intervals between the consumption of milk products and the consumption of meat products.

Fowl

Most domestic breeds of birds are acceptable. Those not acceptable include (1) birds of prey, (2) those having a front toe, (3) those having a craw, and (4) those catching food thrown into the air and then tearing it up on the ground.

Fish

In general, fish with removable scales and fins are acceptable. Sharks and shellfish are not acceptable. There has been some controversy concerning the acceptability of swordfish and sturgeon. Fish that are acceptable usually have the distinction of being pareve or neutral. Fish can be eaten with either a dairy meal or a meat meal, although they generally are not mixed with meat.

The kosher dietary laws are somewhat complex, and both the food processor and the foodservice operator should obtain the correct information from the proper kosher certifying organization to be sure that they are adhering to these laws.

Halal

Like most religions, Islam makes prescriptions about food. Halal is a Muslim word referring to what is lawful or permitted. It is used particularly with food and then relative to meat, meat from animals that have been ritually slaughtered.

The meat of swine is prohibited as is the consumption of blood and most scavenger animals.

Ritual slaughtering requires consecration of the kill by the saying of certain words and the cutting of the throat with one thrust severing both the windpipe and the jugular vein. The blood must be drained.

Game is halal if the words of consecration are spoken when it is shot or when a trained dog is released to retrieve it. Fish are halal if caught when alive. Dead, gathered fish are not.

There is an injunction in Islam to be reasonable in all things. Hence, authorities would probably approve the eating of prohibited food if no alternate to survival existed. "Necessity makes prohibited things permissable," a legal principle. For example, Muslim troops in World War II were known to eat canned ham.

Halal and noningestion of swine are important to Muslims. The Sepoy Mutiny in 1857 by Muslim troops in India was caused by rumors that the bullets were greased with lard. Use of the bullets required the Sepoys to bite the ends of the cartridge patches. The mutiny led to the British government taking over the management of Indian Affairs from the East India Company.

Consult a recent *Food Technology* issue for additional details on halal and information on additional religions (96).

References

1. Cincinnati Enquirer, 13 May 1973.
2. Reichman, M.E. et al. 1993. J. of Nat. Cancer Inst., *89*(9), 722.
3. Latta, S. 1990. Inform *1*(4), 238.
4. *1993 Heart and Stroke Facts*, Washington, DC: American Heart Association. (Issued annually.)

5. *Cancer Facts and Figures, 1993.* Washington, DC: American Cancer Society. (Issued annually).

6. Stamler, J. 1993. Arch. Intern. Med. *153*(9), 1040.

7. Denke, M.A. et al. 1993. Arch. Intern. Med. *153*(9), 1093.

8. Time Magazine, August 1993, p. 56.

9. Lightman, S.W. et al. 1992. N. Engl. J. Med. *327*(27), 1893.

10. Berdanier, C.D. and McIntosh, M.K. 1991. Nutrition Today, *26*(5), 6.

11. Must, A. et al. 1992. N. Engl. J. Med. *327*(19), 1350.

12. Folsom, A.R. 1993. J. Am. Med. Assoc. *269*(4), 483.

13. Wing, R.R. et al. 1992. Am. J. Clin. Nutr. *55*(6), 1086.

14. Inform, *3*(8), 869 (1992).

15. Pediatrics *90*(3), 469 (1992).

16. Inform *1*(6), 525 (1990).

17. Kant, A.K. et al. 1993. Am. J. Clin. Nutr. *57*(3), 434.

18. Food, Fats, and Oils, 7th ed. 1994. Washington, DC: Institute of Shortening and Edible Oils.

19. Knight News Service, Cincinnati Enquirer, May 1991. Information authenticated by Dr. Paul W. Moe in a private communication.

20. Mertz, W. et al. 1991. Am. J. Clin. Nutr. *54*(2), 291.

21. Committee on Diet and Health Food and Nutrition Board. 1989. *Diet and Health.* Washington, DC: National Academy Press.

22. Hegsted, D.M. et al. 1993. Am. J. Clin. Nutr. *57*(6), 875.

23. Hunninglake, D.E. et al. 1993. N. Engl. J. Med. *328*(17), 1213.

24. Food Chemical News, 5 April 1993, 38; Inform *4*(6), 673 (1993).

25. Grundy, S.M. 1986. N. Engl. J. Med. *314*(12), 745.

26. Wardlaw, G. Healthline, 14 May 1989.

27. Marsic, J. and Yodice, R. 1992. Inform *3*(6), 681.

28. Mata, P. et al. 1992. Am. J. Clin. Nutr. *56*(1), 77.

Berry, E.M. et al. 1992. Am. J. Clin. Nutr. *56*(2), 394;

Wahrburg, V. et al. 1992. Am. J. Clin. Nutr. *56*(4), 678.

29. James, P. et al. 1992. Nutr. Reviews *50*, 68.

30. Mensink, R.P. and Katan, M.B. 1990. N. Engl. J. Med. *323*(7), 439.

31. Hunter, J.E. and Applewaite, T.H. 1991. Am. J. Clin. Nutr. *54*(2), 363.

32. Inform *2* (7), 572 (1991).

33. Zock, P.L. and Katan, M.B. 1992. J. Lipid Res. *33*, 399; Food Chemical News 31 August 1992, 3; Troisi, R. et al. 1992. Am. J. Clin. Nutr. *56*(6), 1019; Katan, M.B. and Mensink, R.P. 1992. Nutr. Rev. *50*(4), 46; Mensink, R.P. et al. 1992. J. Lipid Res. 33, (10), 1493.

34. Haumann, B.F. 1993. Inform *4*(5), 562.

35. Inform *2*(7), 578 (1991).

36. Ross Laboratories is a division of Abbott Laboratories (Chicago), located in Columbus, OH.

37. Haumann, B.F. 1993. Inform 4(7), 800.

38. Stamper, M.J. et al. 1993. N. Engl. J. Med. 328(20), 1444.

39. Rimm, E.B. et al. 1993. N. Engl. J. Med. 328(20), 1450.

40. Diplock, A.T. 1992. Inform, 3(11), 1214.

41. Inform 4(8), 934 (1993).

42. Inform 3(12), 1334 (1992).

43. Inform 4(2), 189 (1993).

44. Inform 3(9), 1037 (1992).

45. Lesko, S.M. 1993. J. Am. Med. Assoc. 269(8), 998.

46. Cincinnati Enquirer, 18 November 1992.

47. Hulley, S.B. et al. 1992. Circulation 86(3), 1026.

48. Johnson, C.J. et al. 1993. J. Am. Med. Assoc. 269(23), 3002;
Sempros, C.T. et al. J. Am. Med. Assoc. 269(23), 3009.

49. Krommal, R.A. et al. 1993. Arch. Intern. Med. 153(9), 1065.

50. Cincinnati Enquirer, 20 October 1992.

51. Hulley, S.B. et al. 1993. J. Am. Med. Assoc. 269(11), 1416.

52. Lawn, R.M. 1992. Sci. Am. 266, (6), 54;
Scanu, A.M. 1992. J. Am. Med. Assoc. 267(24), 3326.

53. Nestel, P.N. 1992. J. Lipid Res. 33(7), 1029.

54. Manninen, J. et àl. 1992. Circulation 85(1), 37.

55. Inform 3(4), 459 (1992).

56. Inform 3(12), 1335 (1992).

57. Inform. 1993. 4(4), 354, 364.

58. Willett, W.C. et al. 1992. J. Am. Med. Assoc. 268(15), 2037.

59. Skukla, V.K.S. 1990. Inform 1(2), 150.

60. Cincinnati Enquirer, 19 November 1992.

61. Moran, J.P. et al. 1993. Am. J. Clin. Nutr. 57(2), 213.

62. Trout, D.L. 1991. Am. J. Clin. Nutr. 53(1), 322S.

63. Hunter, J.E. 1991. Inform 2(8), 723.

64. Am. J. Clin. Nutr. 56(6), 1992.

65. Sanders, M.E., Wasserman, B., and Foegeding, E. A. 1993. Food Technol. 47(3), 18S.

66. Kareiva, P. 1993. Nature 363, 580.

67. Quinlan, P. and Moore, S. 1993. Inform 4(5), 580.

68. Public Health Service, 1989. *Promoting Health/Preventing Disease: Objectives for the Nation*, Washington, DC: U.S. Government Printing Office.

69. Yackel, W.C. and Cox, C. 1992. Food Technol. 46(6), 146.

70. Orthoefer, F. and McCaskill, D. 1992. Inform 3(12), 1270.

71. Morrison, R.M. 1992. Inform 3(12), 1270.

72. Bennett, C.J. 1992. Cereal Foods World, 37(6), 429.

73. Thayer, A.M. Chem. and Eng. News, 15 July 1992. 26.

74. Calorie Council Control, 1992 Survey.

75. Cincinnati Enquirer, 3 October 1991.

76. Inglett, G.E. and Grisamore, S.B., 1991, Food Technol, 45(6), 104 and Penichter, K.A. and McGinley, E.J., 1991. Food Technol 45(6), 105.

77. Norris, M.E. and Singras, L.G. 1990. Inform 1(4), 388.

78. Bath, D.E. et al. 1992. Cereal Foods World, 37(7), 495.

79. Hassel, C.A. et al., 1993. Cereal Foods World 38(3), 142.

80. Haumann, B.F. 1993. Inform 4(4), 344.

81. Caragay, A.B. 1992. Food Technol. 46(4), 65.

82. Food Technol., 46(7), 64 (1992).

83. Cincinnati Enquirer, 24 February 1993.

84. Grundy, S.M. 1990. Nutr. Health News 7(2), 2.

85. Cincinnati Enquirer, 10 February 1991.

86. Philadelphia Inquirer (AP), 9 November 1992.

87. Inform, 3(7), 785, 1992.

88. Mermelstein, M.H. 1993. Food Technol., 47(2), 81.

89. The Referee 17(2), 1, 10 (1993).

90. FDA Consumer, an FDA Consumer Special Report, May, 1993.

91. Mermelstein, N.H. 1994. Food Technol., 48(7), 62.

92. Haumann, E.F. 1993. Inform 4(5), 562.

93. Mermelstein, M.H. 1993. Food Technol, 47(4), 65.

94. Kurtzweil, P. 1994. FDA Consumer April, 15.

95. Regenstein J.M. and Regenstein, C.E. 1992. Food Technol, 46(10), 122.

96. Chaudry, M.M. 1992. Food Technol, 46(10), 92.

Chapter 14

Analytical Methods for Oils and Fats

Fred J. Baur, Ph.D.

INTRODUCTION

Analyses of raw materials or finished products are necessary to determine if one has actually obtained what is called for by specifications and standards. This chapter covers: (1) finished oil and fat products, (2) the use of these finished products in food preparation by frying, and (3) in-process or manufacturing checks of the oil and fat products. As the selection of and confidence in the selected method(s) are of vital importance, brief comments on value or purpose are given on each method mentioned.

For complete information on methods for identification of oils and fats and suspected contamination, the reader is directed to *Bailey's Industrial Oil and Fat Products*, Vol. 2, 4th ed., by D. Swern (John Wiley and Sons, Inc., 1982). The bibliography lists additional texts that may be consulted on other aspects not covered.

LEGISLATION AND REGULATIONS

The U.S. Food and Drug Administration (FDA) has established no specific regulations to control oils and fats or their uses. The Canadian counterpart to the FDA, the Health and Protection Branch, has regulations on oils; namely, specific gravity, refractive index, saponification value, acid value, and iodine value. These analyses tend to characterize an oil or fat. Although the FDA has not established standards of identity for edible oils and fats, the agency enforces accurate labeling of oils and fats under the general provisions of the Federal Food, Drug and Cosmetic Act and accepts Codex standards in its regulatory activities.

In 1985 the U.S. Department of Agriculture (USDA) established guidelines for frying oils in factories producing fried meats. Their manual states "large amounts of sediments and free fatty acid content in excess of 2% are usual indications that frying fats are unwholesome and require reconditioning or replacement." Free fatty acid (FFA) is not always a good guideline for frying fats. Two percent seems to be a good maximum for large processors of fried meat and poultry with 1+% for potato chippers. In foodservice, FFA may go higher than about 2% without a problem. The USDA guidelines are the extent of known regulations for the North American continent. The USDA is the U.S. coordinator of activities relating to U.S. participation in the Codex Alimentarius Commission (Codex standards).

A 1991 U.S. survey on regulations to control frying fats turned up no specific regulations except ones ensuring that oils and fats used in foodservice establishments must be obtained from an approved source and not be adulterated. (1) Twenty-one states and 20 major cities were involved. In this same effort, 25 countries throughout the world, including 16 in Europe, were also contacted. Several European countries have specific regulations and these restrict the levels of polar compounds in frying fats (these are degradation products—see the section "Product Use Analyses").

Although formal regulations are uncommon and may remain so, increased awareness of the nutritional aspects of frying fats and fried foods and expanded use of guidelines, new rapid tests, and on-site monitoring devices can be expected to improve the quality of commercially fried foods.

The long-established food laws and regulations are intended to protect consumers and to ensure that foods and food preparation environments are sanitary and free of filth, chemicals, and other contaminants. These are not expected to lessen.

The GRAS (generally recognized as safe, by FDA) affirmed list contains canola oil and completely hydrogenated rapeseed oil. Methods are mentioned. The key comment is that these oils should not contain more than 2% erucic acid (see Chapter 13).

The Food Chemical Codex (FCC) was formed in 1961 in response to a need expressed by the FDA, the public, and the food industry. Its purpose is to define food-grade qualities for the identity and purity of chemicals used in food products. In the United States, the FDA adopts many of the FCC specifications as its legal basis for food-grade quality of food chemicals. In Canada and many other countries, FCC specifications have legal status—they are automatically adopted as stipulated in their respective food laws. To harmonize purity specifications worldwide and to facilitate the international trade of food chemicals, the Committee on Food Chemical Codex works closely with a number of international organizations by communicating its intended specifications. One such organization is the Codex Alimentarius Commission, which was established in 1962 to implement the United Nations Joint Food and Agriculture Organization/World Health Organization (FAO/WHO) Food Standards Program. The Codex Alimentarius Commission also develops specifications for food chemicals by convening an independent panel of experts called the Joint FAO/WHO Expert Committee on Food Additives.

Specifications for food chemicals are based on strict identity and purity standards. In commerce, they often serve as a basis for buyer–seller agreements between food chemical manufacturers and food processors and as a means for compliance as requested by national authorities responsible for regulating foods and food chemicals.

FCC specifications are published in the form of monographs written for a specific food chemical or a group of related food chemicals. Each monograph contains specifications for the major constituent of the food chemical, as well as other appropriate specifications that define the material, for example, those that limit the presence of impurities such as lead, heavy metals, and arsenic. Each specification includes a limit as well as a corresponding test method that enables food chemical manufacturers and customers to show compliance with the limit. In total, the assay and purity specifications provide sufficient information to define food-grade quality for a food chemical.

The first edition of FCC was published in 1966. The third supplement to the third edition was published in 1992 (2). When the fourth edition is released, now expected in 1996, in the vicinity of 1000 food chemicals will be included.

An example of a monograph in the oil and fat area is for unhydrogenated soybean oil as listed in the second supplement to the third edition. The fatty acid composition profile, in percentages, is listed as follows: <Cl4, <0.1; 14:0, <0.5; 16:0, 7–12; 16:1, <0.5, 18:0, 2–5.5; 18:1, 19–30; 18:2, 48–58; 18:3, 5–9; 20:0, <1.0; and 20:1, <1.0. The tests and their limits are arsenic—not more than 0.5 mg/kg; cold test—passes test; color (Lovibond)—not more than 20 yellow/2.0 red; FFA (as oleic)—not more than 1.0%; iodine value—between 120 and 143; lead—not more than 0.1 mg/kg; peroxide value—not more than 10 meg/kg; stability (active oxygen method)—not less than 7 h; unsaponifiable matter—not more than 1.5%; water—not more than 0.1%.

Monograph revisions and additions are an ongoing process reflecting volume of use of the chemical, recognition of changes in safety concerns, changes in methodology, and so on.

AOAC INTERNATIONAL

AOAC International is the premier organization worldwide in providing validated methods for food analyses and a leader in methodology for food oils and fats. Until August 1991, the letters AOAC stood for the Association of Official Analytical Chemists.

AOAC International (AOAC) is an independent association devoted to the development, testing, validation, and publication of accurate and precise methods of analyses of commodities related to agriculture and public health and safety.

The main reason for the primacy of the AOAC is that various regulatory agencies at the international, national, and local levels apply AOAC methods as the methods of choice in their monitoring commodities and industries under their surveillance; that is, AOAC methods are preferred for the regulatory process.

A bit of history will explain this development. The AOAC was formed in 1884 as the Association of Official Agricultural Chemists and made part of the United States Department of Agriculture (USDA). This came about as the result of a meeting of state agricultural officials and their recognition of the need to adopt uniform methods of analyses for fertilizers. By 1887, AOAC was also publishing methods of analyses for feeds and dairy products. Passage of the U.S. Pure Food and Drug Act in 1906 saw the AOAC entering many other food areas. The AOAC remained in the USDA until 1927 when the Food and Drug Administration was created. The name was changed to Analytical Chemists from Agricultural Chemists in 1965 and International was added in 1991. The most significant change occurred in 1979 when the AOAC became independent of the FDA while still relying on the FDA as the key source of dollars (grant support) and volunteers. FDA to this day remains a strong contributor of dollars and volunteers. In the past decade, industry, where it had been welcomed aboard via lower echelon volunteers, has gradually become a partner to agencies and academia in running the association. The AOAC now truly belongs to the entire concerned scientific community.

AOAC at this time maintains no laboratories, conducts no research, and performs no tests. The actual work of devising and testing methods is done by members of the association in their official and professional capabilities as staff scientists of various regulatory agencies, experiment stations, colleges and universities, commercial firms, and consulting laboratories.

Regulatory use of AOAC methods practically commands knowledge of the methods and the system for their development and approval. The AOAC system or protocol, variously called collaborative study, methods validation, the referee system, is second to none in thoroughness, and it is cost-shared. As a minimum a protocol (study design) is submitted by the topic associate referee (examples include hydrogenated fats, oxidized fats) to his or her general referee (fats and oils), the Methods Committee chairperson (Food Nutrition), and the committee statistician before the study is started. In addition, before starting the study, the associate referee should test the ruggedness of the method to determine the tolerance of the method to inadvertent or intentional deviations from the written procedure in such operational steps as reaction time, amount of reagent, and so on. All studies require at least six collaborators (laboratories) and five samples to yield a minimum of 30 data points. AOAC-approved methods are quality methods, considered accurate, reproducible, and proven.

Official methods are described as first action or final action, and, in a few cases, procedures. A valid criticism of the approval process is the time required. To take a developed method through approval to first action has been estimated as requiring 4–9 months, depending on the type of method. Methods may be recommended for final action after they have been published in first action for at least 2 years.

The most recent edition of *Official Methods of Analyses*, the 15th edition, was published in January 1990. Updates are released annually and are included in the purchase price.

OTHER INVOLVED/CONCERNED METHODS ORGANIZATIONS

There are two other organizations of note in methods development and approval relative to oils and fats.

The first is the American Oil Chemists Society (AOCS). It is the key technical and methods development and approval organization worldwide in the oils and fats area. The AOCS has standardized methodology for more than 75 years. The fourth edition of the "AOCS Official Methods and Recommended Practices" was released in December 1989. Annual additions and revisions are published.

The number of AOCS approved methods far exceeds the number of AOAC approved methods for oils and fats. This is understandable because the regulatory agencies are concerned with health, safety, and cleanliness and have no official responsibilities for key quality attributes for the consumer such as appearance, color, flavor, and texture. The agencies' prime interest is in finished products, with lesser interest in raw materials to no interest in process checks. Agencies are, of course, vitally interested in possible contamination and/or adulteration of raw materials.

The interaction/cooperation between the AOCS and AOAC over the decades has been excellent. A key example was the aflatoxin problem in the early 1960s (3). In many instances and certainly those of regulatory interest, AOCS will use the AOAC protocol. More uniform statistical handling remains a goal.

Since 1985, the AOCS has been working with both national and international organizations to harmonize analytical methods, to make uniform the methods used in international trade of oils and fats (4).

In particular this has meant interacting with the Federation of Oils, Seeds and Fats (FOSFA). FOSFA International has a lead role in providing a forum for trade associations and organizations that set technical standards to review, study, and harmonize oil and fat methodology. FOSFA has recently become the secretariat for the International Standards Organization (ISO) Commission on Oils and Fats.

The second organization of note is the International Union of Pure and Applied Chemistry (IUPAC). For at least 20–30 years the AOAC and IUPAC have been cooperating on methodology for commodities of common interest. This cooperation has recently widened to include technical societies like the AOCS.

The American Association of Cereal Chemists (AACC) is another concerned technical association with standards and approved analytical methodology. The AACC relies on the AOCS for its methodology in the oil and fat area.

The Institute of Food Technologists (IFT), the dominant food technical association in the world and one that is highly research oriented, has no standards/methods program.

Like the AOAC, none of the organizations listed above possesses laboratories with staff to conduct methods research and/or testing.

The American Society for Testing and Materials (ASTM) is a minor factor in the oil and fat area.

SPECIFICATIONS/STANDARDS

A specification is a clear and accurate description of the technical requirements for materials and products including minimum requirements of quality necessary for an acceptable product. One needs to determine what one's expectations or requirements are for the products to be purchased and then phrase these in the language of the trade with definitions of terms as may be required by the compiler and/or user.

These specifications are necessary to ensure that the buyer is receiving what is desired, what is being paid for. Finished product specifications must ensure quality is maintained for continuing success in the market arena.

Specifications drawn up for any material or product become the standards for that material.

Standards are protocols, the ways of doing something, and for finished products, they are the indicators of quality. Depending on the comparative importance and possible regulatory interest, the standards may be obligatory or optional. An example of obligatory standards is standards of identity. Standards can be exclusive, that is, no ingredient not specifically included in the standard may be present.

A couple of examples using methods will be of value.

In the purchase of a crude oil it is likely that "limits" will be set for appearance, aroma, chlorophyl, color, free fatty acids, flavor, and unsaponifiables. Depending on the intended end use, others can be added like chill test, filter test, phosphorus, and soap. The required absence of substances can be mentioned; for example, no anti-oxidants.

Finished products will have similar listings. Common are color, flavor, free fatty acids, and moisture. If the product is gas (nitrogen) packed, dissolved nitrogen is likely. If the product is a shortening, consistency or penetration is important.

FINISHED PRODUCT ASSAYS

This section lists a number of methods, many of which have been used for finished product assays, both oils and shortenings. Those associations having standard methods are noted.

AOM (Active Oxygen Method)—AOCS. A predictive, accelerated rancidity test; a measure of oxidative stability. It is not very accurate. It has more value in shelf-life estimation of baked products than fry life. The AOM is being replaced by the oil stability index (OSI), which has greater precision. The OSI is automated using instrumentation by Omnion, Inc. (Oxidative Stability Instrument) or Brinkmann Instruments (Rancimat). The collaborative study was completed in 1991. Other methods to predict stability include the Kreis test, O_2 bomb, and Swift. The Japan Oil Chemists' Society has recently developed a conductimetric procedure to serve as an alternate to the AOM.

Color—AOCS. The Lovibond Tintometer is most generally used, a must method. After all, the appearance of a product is its first hurdle in consumer acceptance. Red and yellow are the colors usually measured.

Consistency—AOCS, ASTM. See Penetration.

Dropping Point (Slipping Point)—AOCS. Melting measurement frequently used by European laboratories.

FFA (Free Fatty Acids)—AOCS, AOAC. An indicator of quality, the freshness of the fat, the efficiency of the refining process. Not a flavor predictor. FFA and acid value are the same measurement but the calculations differ. Obviously, the higher the free fatty acids in a crude oil the greater the refining loss.

Flavor—AOCS. Probably the single most important quality factor. Organoleptic testing is common. Gas-liquid chromatography (GLC) can be used on headspace and on the oil or fat. See Chapter 16.

Hexane—AOCS. Measures residual oil extraction solvent by GLC.

Iodine Value—AOCS, AOAC, IUPAC. A measure of the number of double bonds or the degree of unsaturation and, therefore, an indicator of oxidative stability. Not a good indicator of fry life. Hanus and Wijs are the common reagents used. Wijs is said to have a bias. The AOCS has replaced CCl_4 (carcinogen) by cyclohexane or cyclohexane-acetic acid (1:1) as solvents. Iodine value is now commonly calculated from GC-FAC with equivalent accuracy and precision and without the need for the highly hazardous chemicals in the titration methods.

Melting Point—AOCS, AOAC. The point at which a fat becomes completely liquid and clear. It is of comparatively little value, but still used. It gives no real clue to the solids present at lower temperatures. Melting points of typical fats and oils are rarely sharp because they contain a mixture of triglycerides comprised of fatty acids of varying chain lengths and unsaturations. Polymorphic forms will also influence the melting point. See Wiley Melting Point for information on preferred methods.

Moisture—AOCS, AOAC. Moisture, and moisture and volatile matter methods abound. The AOCS has at least 17 on oils and fats. Most natural products are likely to have some water present, and oils and fats are no exception. Rarely are products completely dry. They can take up moisture just on standing, hence, it is almost impossible to keep a product dry. Water solubility in oils and fats runs 0.05–0.30%. Karl Fischer automatic titrator is the moisture procedure of choice. It is rapid, precise, and no heat is required. Moisture analyses are important in trading.

% Monoglycerides—AOCS, AOAC, IUPAC. Superglycerinated fats, which are high in monoglycerides, are the most common emulsifier used in shortenings. They improve baking performance. The four current approaches to monoglyceride analysis involve: titration, column chromatography, capillary GC, and HPLC. The commonly used titrametric method uses periodic acid to determine alpha-monoglycerides but involves chloroform as a solvent. The AOCS procedure has two parts: section I for monoglyceride contents below 15% and, therefore, on shortenings, and 45% and above for the concentrates. The probability is that this method will be kept after modification.

The column chromatographic procedure (AOAC) separates monoglycerides, diglycerides, and triglycerides. Modification has been proposed to substitute ethyl ether/petroleum ether for benzene, by the AOCS. The AOCS has adopted the IUPAC capillary GC but the method lacks sample flexibility. At the time of this writing, there were no approved or official HPLC methods. A collaborative study is underway on a HPLC-ELSD (Evaporative Light-Scattering Detection) method. The Uniform Methods Committee of the AOAC International has recommended the adoption of a method that can determine the major neutral lipid classes to replace the recently adopted IUPAC monoglycerides method. In addition to the normal phase liquid chromatography with evaporative, light-scattering detection, gas chromatography and supercritical fluid chromatography should be considered. The separated lipid classes are to include free fatty acids, steryl esters, and monoglycerides, diglycerides, and triglycerides. A recent *Inform* issue has a brief review of the methodology (5).

OSI (Oxygen Stability Index)—(AOCS). An automated replacement for the AOM. It measures the point of maximum change of the rate of oxidation; broad applicability including crude oils.

Penetration—AOCS, ASTM. A measurement of the firmness, the hardness, and the resistance to penetration. The plasticity and the plastic ranges of shortenings and margarines are an important quality attribute. A penetrometer is used with the ASTM one being the standard. Formulation and processing are the keys to consistency, or penetration, which varies with temperature.

Peroxide Value (PV)—AOCS, AOAC. A measure of the state of oxidation of an oil or fat. Influencing factors are the constituent fatty acids and the length and type of storage. It is of little value in examining frying fats as the peroxides are lost during frying.

Reichert-Meissl, Polenske, and Kirschner Numbers—AOCS, AOAC. These are methods for the lower-chain saturated fatty acids: Reichert-Meissl for volatile, water soluble, or C_4 and C_6; Polenske for volatile, water insoluble, or C_8, C_{10}, and C_{12}; and Kirschner for C_4.

Saponification Value (SV)—AOCS, AOAC. An indicator of molecular weight or size as a function of the chain lengths of the constituent fatty acids. High levels of unsaponifiables throw off the interpretation. It is of comparatively little value for finished product control.

Smoke Point, Fire and Flash Points—AOCS. An indicator of the nontriglyceride components such as free fatty acids, monoglycerides, diglycerides, glycerol, and hexane; not pertinent to fry life. Low smoke points may result in low frying temperatures, particularly in the home.

Solids: Solids Fat Index (SFI)—AOCS; Solid Fat Contents (SFC)—AOCS, IUPAC. The measurement of the solids over a range of temperatures is a key in predicting plastic range and plasticity. Dilatometry (SFI), the measurement of volume changes as a function of temperature is being replaced by low-resolution magnetic resonance (SFC). The reasons are (1) greater accuracy, which is needed in new product development, (2) increased speed, (3) lower analytical costs (once the instrument has been purchased), and (4) dilatometry is not suitable for on-line measurements.

An international collaborative study is now underway comparing the Indirect Method (SFC), which compares the magnetic resonance signal of the liquid in the sample to a reference oil, and the Direct Method, which uses signals from both the liquid and solid in the sample (6). The study is expected to lead to the recommendation that the Indirect Method be made the standard one. At this time, the Indirect Method (AOCS) is used in the United States and Japan, whereas the Direct Method (IUPAC) is used in Europe.

Totox Value. The Totox Value (Total Oxidation Value) measures oil deterioration and is an addition of $2 \times$ PV plus the anisidine value (see Chapter 16). It is used in Europe.

Unsaponifiables (Unsap)—AOCS, AOAC. The usual unsaponifiables are hydrocarbons and higher alcohols including sterols. A possible hydrocarbon adulterant is mineral oil.

Wiley Melting Point—AOCS, AOAC. The Mettler Dropping Point Instrument has replaced the Wiley as an aid in blending and hydrogenation (an indicator of solids). European laboratories frequently use the Wiley. Declared surplus in 1991. It will be removed from the Book of Methods (AOCS) in 1994. A recent survey (7) of 565 methods users on which melting point methods are used yielded the following: Mettler 60, open capillary 56, Wiley 35, closed capillary 34, "slippoint" 4, and flow test 1. (Forty-two percent of the 165 responders said none used.)

PRODUCT USE ANALYSES

This section deals principally with deep fat frying. See Chapter 7 for an earlier discussion of the chemical reactions and resulting products.

Frying

Introduction

Deep fat frying has by far received the most attention relative to changes in oils and fats during product use. The two important stimuli for the considerable efforts are: (1) concern with safety/nutritive value—the possible formation of toxic or deleterious materials as a consequence of the exposure to heat and oxygen—and (2) the changes in the frying medium that influence the sensory quality of the oil or fat and the food fried in it.

The number of changes that take place in oils and fats during heating in deep fat frying involve a complex pattern of thermolytic and oxidative reactions. The reactions are generally considered to be (1) hydrolytic, (2) oxidative, and (3) polymeric. The resulting products are of great interest and importance to the food industry, as they can and will impact consumer acceptability.

Some of the reaction products are volatile. These volatile decomposition products include hydrocarbons, aldehydes, ketones, furans, and carboxylic acids. These materials are of interest as (1) they are indicative of the chemical reactions taking place during frying, (2) they are inhaled by the operators of the fryers, (3) some do remain in the oil and by entering the fried product are ingested, and (4) as flavor is greatly influenced by odor, they contribute to the organoleptic acceptability of the fried food.

The nonvolatile decomposition products are polar and nonpolar. These include cyclic monomers, non-cyclic monomers, dimers, trimers, and higher-molecular-weight compounds. Obviously, they remain with the oil and are absorbed by the fried food and eaten by the consumers. They are reliable indicators of fat abuse because their accumulation is steady, being nonvolatile. These reaction products are responsible for the physical changes in the frying medium—increases in viscosity, color, and foaming—and the chemical changes such as increases in FFA, carbonyl value, hydroxyl content, SV, and decreases in unsaturation and, ultimately, increases in the formation of high-molecular-weight products.

Much attention has been given to the polar compounds or materials. Polar materials include FFA, monoglycerides and diglycerides, monomers, dimers, polymers, oxidized fatty acids, soaps, browning reaction products, and other trace materials like metals.

Polar materials can increase from less than 5% level to greater than 25% during foodservice frying operations.

Methods

Most methods for measuring deterioration of frying fats are based on changes in the nonvolatile decomposition products.

Traditionally, nonspecific methods such as FFA, IV, viscosity, nonurea adducting esters, petroleum ether insolubles, and oxidized fatty acids have been used. None has proved to be a good measure of the condition of an oil in use. The specific PV is also not a good measure because, as has been stated, peroxides are not stable under frying conditions.

Standard methods that have been adopted include:

- Polar compounds—AOAC, IUPAC, AOCS; via column chromatography; simple, accurate, reproducible
- Conjugated dienoic acids—AOAC, IUPAC; via ultraviolet absorption; levels plateau on continued use, probably because the dienes form polymers
- Fatty acid analyses and C18:2/16:0 ratio—AOCS; via gas-liquid chromatography
- Polymerized triglycerides—IUPAC, AOCS (to be adopted by AOAC); via gel permeation, HPLC

As the above methods are somewhat time-consuming, the polar compound method requiring 3.5 h for a complete run, a number of commercially available quick tests have become available (8). The following is a list of what include and what they measure:

- dielectric constant measured by the Food Oil Sensor (FOS)—polar molecules.

- Oxifrit, formerly RAU-Test (colorimetric)—oxidized compounds.
- Fritest (colorimetric)—carbonyl compounds.
- Spot Test (colorimetric)—FFA.
- alkaline contaminant materials (colorimetric)—soaps.
- Veri-fry—total polar, FFA, total alkaline.

A comparison of the first four tests with the polar compound analysis showed the FOS to have the best correlation, perhaps because they measure the same materials. Another investigator claims that the FOS is most convenient for quality control purposes (9). It is said to be influenced by outside factors, such as water or fat extracted from the food, hence requiring standardization each time it is used. It also has been suggested that an FOS reading of 4.0 is indicative that the used fat should be discarded. The AOAC associate referee for oxidized fats (M. M. Blumenthal) is planning a collaborated study with several of the recently developed quick test kits (10).

Other methods that have received investigation include:

- Gas liquid chromatography for
    Oxidative and thermal dimers
    Dimeric triglycerides
    Cyclic monomers
    Total polymers
- Size exclusion chromatography for polar artifacts
- Gel permeation chromatography for dimeric and oligomeric triglycerides
- Size exclusion chromatography for triglycerides, dimeric triglycerides, and polymers

Methods such as smoke point and epoxy values have not recently received much attention. Some investigators have used a combination of methods to obtain greater reliability.

Any method chosen for use should depend on the need or purpose of the measurement. Is it for quality control or research? What are the needed accuracy and reproducibility? What speed is required? What are the costs of establishing and then running the procedure?

Baking

The only methods found for fat or lipids relative to baked products have been issued by the AOAC. The 15th edition of the Official

Methods contains eight methods including fat and fat number in bread, lipids in flour, and fat in flour. This last method, 922.06, is also recommended for fat in bread and in baked products. One last example: fig bars and raisin-filled crackers have their own method.

NUTRITIONALLY IMPORTANT ANALYSES

Chapter 13 deals in some detail with not only the nutritional importance of fat level but also the fat type. The fat type is primarily a function of the constituent fatty acids.

This section covers methods, most not standard, that deal with fatty acids.

Table 14–1 lists the information (11).

As of this writing, trans acids remain one of the more important nutritional questions on oils and fats. An update by the General Referee of the AOAC for Fats and Oils should be of interest. "Linoleic acid isomers (c-c/c-t/t-c/tt) are adequately separated by capillary GC, whereas trans-monoene isomers are not all separated from cis-isomers." Thus Ratnayake, et al. (12) proposed use of a combined capillary GC and IR method for determination of cis and trans-18:1 fatty acids in partially hydrogenated oils. [The method is to be collaboratively studied in a joint study carried out by Ratnayake (13).] For further information consult Firestone's chapter in a recently published book (14), and two recent issues of Inform (15, 16).

FAT ANALYSES (AOAC) FOR NUTRITIONAL LABELING

The AOAC-International recently established a Task Force on Nutrient Labeling with the responsibility of assessing adequacy of the Official Methods of Analyses for this purpose. Early in this effort it was learned that laboratories using official AOAC methods for fat determination are concerned with the appropriateness of the methods for various food matrixes, the lack of official methods for some matrixes, and the absence of a clear definition of fat for labeling purposes. A subcommittee was, therefore, formed to address these three concerns.

The methods available for fat determination vary in solvents used and in whether and how samples are hydrolyzed or digested prior

Table 14.1. Nutritionally significant analytical methods.

Technique	Standard or official	What measured	"Value" comments
Argentation thin-layer chromatography	—	Monenes and dienes	Separates cis and trans enes; valuable adjunct to gas liquid chromatography
Infrared spectrophotometry	AOCS, AOAC	Trans acids	Best trans measurement for routine analyses; can be applied directly, no derivatives needed; bias below 15%
Gas chromatography	AOCS, AOAC, IUPAC	Fatty acid analyses	The method for this need; via methyl esters; % cis-cis
Gas-liquid chromatography	—	Trans acids	More accurate than IR; provides more information; trans on both monenes and dienes; incomplete separation of 9, 12, 18:2 isomers
High-performance liquid chromatography	—	Resolves positional isomers/polymers	A complement to GLC

Method	Standards	Application	Comments
HP size exclusion			State of polymerization and/or amount of use
Lipoxidase	AOCS	Specific for cis-cis methylene interrupted	For essential fatty acids
Liquid chromatography (reversed phase)	AOAC, IUPAC AOCS, AOAC AOCS, AOAC, IUPAC	Free tocopherols Polymerized triglycerides N-3 and N-6 unsaturated fatty acids	
Mass Spectrometry	AOCS, AOAC, IUPAC IUPAC	Unsaturated fatty acids Single fatty acids	Gives molecular weight and number and position of double bonds. Used in conjunction with separation techniques
Natural Abundance NMR	—	Trans isomers Single fatty acids Positions of double bonds	
Ozonolysis	—		Prior conversion and separation needed
% Tocopherols	AOCS	Level of Vitamin E	
Proton Nuclear Magnetic Resonance	—	Geometrical isomers	Used in conjunction with other separational techniques. Can be applied to glycerides
Ultra violet spectrometry	AOCS, AOAC	Conjugated dienes and trienes	Does not distinguish positional and geometric isomers

to extraction. Thus, it is likely that different components are being extracted in varying amounts and called fat. As an example, acid hydrolysis of high-sugar-containing products prior to ether extraction may lead to anomalously high fat values due to extraction of sugar or sugar by-products. Reextraction may give an accurate fat determination, but this needs verification. Some polysaccharide or protein-containing matrixes may preclude removal of fat with ether, hence requiring a more rigorous procedure, such as hydrolysis, for complete extraction.

In view of some methods inconsistencies, the subcommittee recommended a single concise definition of fat be adopted by the agencies proposing labeling regulations and that AOAC-International validate methods that most accurately measure the food components that are defined as fat.

Fat has been loosely and interchangeably described as lipids (17). The descriptions were attempts to distinguish fat from other nutrients based on its solubility in nonpolar solvents (18). By classical definition, total lipids refer to a sum of triglycerides, diglycerides and monoglycerides, free fatty acids, phospholipids, glycolipids, sphingolipids, terpenes, steroids, sterols, waxes, and other compounds (19). There is little question that glycerides are considered fats (and oils). In addition, sphingolipids, terpenes, steroids, sterols, waxes, fat soluble vitamins, and other compounds are usually not considered dietary fat. Classification of phospholipids and glycolipids continues to be debated. Certainly, free fatty acids must be considered fat from a dietary standpoint.

Currently, fat refers to triglycerides in the United States (20) but total lipids in the European Community (21). Recently, in the United Kingdom it was proposed that total fat intake be calculated as a sum of fatty acids and glycerol corresponding to reporting fat as triglycerides (22).

It was the recommendation of the subcommittee that a single definition of fat, such as the sum of fatty acids from a total lipid extraction, be adopted. A true index of total dietary fatty acids would help consumers judge probable physiological effects of the foods they are consuming. So, from both consumer education and analytical perspectives, a single concise definition of fat will assist accurate and unambiguous label information.

Table 14–2, prepared by the subcommittee covering fat analysis by AOAC methods, has been reproduced with permission of the AOAC.

Table 14.2. Fat analysis for nutritional labeling—AOAC methods.

AOAC no.	Title	Brief description	Applicable matrixes	Comments
983.23	Fat in Foods	Amylase/protease treatment, chloroform-methanol extn	All	Mth ext tot lipids incl mono-, di-triglycerides, steroids, glycolipids, phospholipids and waxes; mth is *not* for determining tot fat. May be used as an extn mth for various lipid fractions
960.39	Crude Fat in Meat	(Pet or diethyl ether) extn	Baby food-meats, meats	Mono-, di-, and triglycerides and traces of other lipid components
976.21	Crude Fat in Meat	Sp gr mth; tetrachloroethylene extn	Baby food-meats, meats	Mono-, di-, and triglycerides, and most of the sterols, glycolipids, phospholipids and waxes
985.15	Crude Fat in Meat & Poultry Prod	Rapid microwave-methylene chloride solv extn	Baby food-meats, meats	Mono-, di-, and triglycerides, glycolipids, phospholipids, and waxes, yield of sterols may be depressed
922.06	Fat in Flour	Acid hydr., pet and diethyl ether extn	Cereals & prod, oils/fats (dressings), potatoes & prod	Mono-, di-, and triglycerides, fatty acid portion of phospholipids and glycolipids; ext nonlipid matl and may overest. fat content; recommend review or study of mth

Table 14.2. Continued Fat analysis for nutritional labeling—AOAC methods.

AOAC no.	Title	Brief description	Applicable matrixes	Comments
920.39B	Crude Fat in Animal Feed	Diethyl ether extn	Cereal & prod	Mono-, di-, and triglycerides and traces of other lipid components
920.39C	Crude Fat in Animal Feed	Diethyl ether extn; H$_2$O prewash if high in sugar	Cereals & prod, sweet mixes (cakes & pies)	Mono-, di-, and triglycerides; may not quant, extract tot lipids; recommend further review or study of mth
945.18A	Fat in Cereal Adjuncts	Pet ether extn	Cereals & prod, sweet mixes (cakes and pies)	Mono-, di-, and triglycerides; may not quant, extract tot lipids; recommend further review or study of mth
925.12	Fat in Macaroni Prod	Acid hydr., pet and diethyl ether extn	Cereals & prod	Mono-, di-, and triglycerides, and fatty acid portion of phospholipids and glycolipids; ext nonlipid matl and may overest. fat content; yield of sterols may be depressed; recommend further review or study of mth
945.38F	Fat in Grains	Refers to 920.39C	Cereals & prod	Mono-, di-, and triglycerides and traces of other lipid components

933.05	Fat in Cheese	Acid hydr., pet and diethyl ether extns	Cheese	Mono-, di-, and triglycerides and traces of other lipid components; yield of sterols greatly reduced
905.02	Fat in Milk	Alk treatment, pet and diethyl ether extns	Dairy	Mono-, di-, and triglycerides and traces of other lipid components
989.05	Fat in Milk	Alk treatment, pet and diethyl ether extn	Dairy	Mono-, di-, and triglycerides and traces of other lipid components
938.06	Fat in Butter	Diethyl or pet ether extn	Butter	Mono-, di-, and triglycerides and traces of other lipid components
920.111A	Fat in Cream	Alk treatment, pet and diethyl ether extns	Dairy	Mono-, di-, and triglycerides and traces of other lipid components
920.111B	Fat in Cream	Babcock, acid hydrolyses, vol. anal	Dairy	Mono-, di-, and triglycerides, phospholipids and reduced sterol yield
952.06	Fat in Ice Cream & Frozen Desserts	Alk treatment, pet and diethyl ether extns	Dairy	Mono-, di-, and triglycerides and traces of other lipid components
945.48G	Fat in Evaporated Milk	Alk treatment, pet and diethyl ether extns (refers to 905.02)	Dairy	Mono-, di-, and triglycerides and traces of other lipid components

Table 14.2. Continued   Fat analysis for nutritional labeling—AOAC methods.

AOAC no.	Title	Brief description	Applicable matrixes	Comments
932.06	Fat in Dried Milk	Alk treatment, pet and diethyl ether extn	Dairy	Mono-, di-, and triglycerides and traces of other lipid components
948.15	Crude Fat in Seafood	Acid hydr. pet and diethyl ether extns	Fish, shellfish	Mono-, di-, and triglycerides, fatty acid portion of phospholipids and glycolipids; in some prod with high sugar content, may overest. fat; may reduce yield of sterols
964.12	Crude Fat in Seafood	Babcock, acid hydr., vol. anal	Fish, shellfish	Mono-, di-, and triglycerides, phospholipids; may reduce yield of sterols
986.25	Fat in Milk-Based Infant Formula	Alk treatment, pet and diethyl ether extns; re-extn (refers to 945.48G)	Infant formula/ medical	Mono-, di-, and triglycerides and traces of other lipid components
925.32	Fat in Eggs	Acid hydr, pet and diethyl ether extns	Eggs/egg prod	Mono-, di-, and triglycerides, fatty acid portion of phospholipids and glycolipids, and may reduce yield of sterols
948.22	Crude Fat in Nuts and Nut Prod	Diethyl ether extn	Nuts	Mono-, di-, and triglycerides and traces of other lipid components

950.54	Tot Fat in Food Dressings	Acid hydr. pet and diethyl ether extns	Oils/fats (dressings)	Mono-, di-, and triglycerides, fatty acid portion of phospholipids and glycolipids
935.39D	Fat in Baked Prod	Acid hydr. pet and diethyl ether extns (refers to 922.06)	Baked cereal prod	Mono-, di-, and triglycerides, fatty acid portion of phospholipids and glycolipids; may reduce yield of sterols; ext nonlipid matl and may overest. fat content; recommend further review or study of mth
945.44	Fat in Fig Bars & Raisin Filled Cookies	Acid hydr. pet and diethyl ether extns	Sweet mixes (cakes & pies)	Mono-, di-, and triglycerides of fatty acid portion of phospholipids and glycolipids; may reduce yield of sterols re-extn; may not remove all sugars
963.15	Fat in Cacao Prod	Pet ether extn	Chocolate prod	Mono-, di-, and triglycerides and traces of other lipid components
925.07	Fat in Cacao Prod	Pet and diethyl ether extns	Candy	Mono-, di-, and triglycerides and traces of other lipid components
920.177	Ether Extract of Confectionery	Pet & diethyl ether, extn; re-extn	Candy	Mono-, di-, and triglycerides and traces of other lipid components

Table 14.2. Continued   Fat analysis for nutritional labeling—AOAC methods.

AOAC no.	Title	Brief description	Applicable matrixes	Comments
920.172	Ether Extract of Prepared Mustard	Diethyl ether extn	Mustard	Mono-, di-, and triglycerides and traces of other lipid components
963.72[a]	Methyl Esters of Fatty Acids in Oils and Fats	GC mth following prep of methyl esters (according to 969.33)	All	Fatty acid profile; recommend further review or study of mth
979.19[a]	Cis, Cis-Methylene Interrupted Polyunsaturated Fatty Acids in Oils	Spectrophotometric mth	All	

This list has been prepared thanks to the efforts of S. Bailey, D. Carpenter, H. Chin, J. DeVries, N. Fraley, W. Hummer, A. Kistler, S. Lee, J. Ngeh-Ngwainbi, P. Oles, and D. Sullivan.
[a]Fatty acid methodology and cis, cis-methylene interrupted procedure.
Source: Reprinted from *The Referee*, Volume 16, Number 69, pages 006–007, 1992. Copyright 1992 by AOAC International.

The AOAC published, also in *the Referee*, updated total fat methods (23) as a result of the publication of the regulations of the Nutritional Labeling and Education Act. In the Act, total fat is defined as "total lipid fatty acids as expressed as triglycerides" and saturated fat as "the sum of all fatty acids containing no double bonds." The Nutrient Labeling Task Force of the AOAC, in view of these definitions, considers none of the AOAC methods to be ideal, but those listed in Table 14.2 other than 983.23, 922.06, 925.12, 964.12, and 935.39D are considered to be adequate. New methods designed to meet the definitions are being pursued.

The AOAC is expected to form a Technical Division on Reference Materials (24). Responsibilities would include (1) providing appropriate reference materials for collaborative studies and routine use of validated methods for analytical measurements in response to the U.S. Nutritional Labeling and Education Act and (2) planning and organizing the international symposium series on Biological and Environmental Reference Materials (BERM).

A task force review was published in (25) *Food Technology* (July 1994) after this chapter was written.

OPERATING METHODS

A number of analytical methods are used to control the processing of oils and fats. These include:

Anisidine value: determines the amounts of secondary and some primary aldehydes derived from peroxides; therefore, a measure of oxidative deterioration

Color (Lovibond)—AOCS: measures efficiency of bleaching and deodorization processes

DSC (Differential Scanning Calorimetry): monitors hard stock levels in blending process

Filter Grade: a check on removal of bleaching earths

Free Fatty Acids—AOCS, AOAC: determine state of oxidation, hydrolysis

Nephelometry: a turbidity measurement to check on refining process

Neutral Oil and Loss—AOCS: to gain insight into anticipated refining loss; hence, an estimate of value of the crude oil

NIRA (Near-Infrared Reflectroscopy): in combination with other measurements like sugars gives an indication of seed quality; a screening tool for oils from seeds

%O$_2$ in Headspace: a check on the efficiency of nitrogen packing on finished oils

Peroxide Value—AOCS, AOAC: a determination of the state of oxidation

Polymer: amount of use or exposure; using HPLC, can be run in 15 min

RI (Refractive Index)—AOCS: a check on the hydrogenation process

Solids—AOCS, IUPAC: measurement of blending and consistency

ON-LINE MEASUREMENTS

The food processing industry has been slow to adopt recent advances in instrumentation and control to on-line measurements (26). Factors mitigating against use of on-line are: (1) the food industry is low profit and automation is expensive and (2) food processes generally are not easy to automate because natural materials show variabilities. Whereas these restraints remain, great strides have been made in the past 1–2 years. Color (27), fat and moisture (28), and hydrogenation (29) are all now feasible. The last can be accomplished by either fiber optic refractometry or Fourier transform infrared spectometry. The latter, in addition to degree of hydrogenation, also gives information on the type of hydrogenation.

PRECISION AND ACCURACY

Precision is the capability of a method to reproduce the results. Accuracy is the ability to yield true values, information on what is actually there.

It is the intent of the AOAC to determine the accuracy of a method whenever possible, whether by the use of standard reference materials or via spiked samples.

It has been only the past few years as new methods are approved, or existent ones revised and approved, that the AOCS and AOAC have been including precision and/or accuracy information in their books on official methods. By far, the majority of approved methods still lack this type of valuable data so that publications reporting on the methods studies have to be consulted.

Accuracy can be very important, particularly if product claims and nutrition are involved. An example in 1991 involved a baking mix and a % fat claim. It turned out that the involved manufacturer was vindicated because the official method was found to have a high bias via picking up oxidized fatty acids from the flour. Be aware that many of such older methods may have similar pitfalls.

SUMMARY

This chapter covers a portion of the methods that can be applied to oils and fats. Included are methods for finished products, the monitoring of their use—particularly in frying—those of nutritional significance, and a brief mention of those that are used for control of processing. It does not cover methods pertinent to identification, the possible presence of toxic or deleterious materials, low levels of natural and added minor components, or lipids as a broad class of materials.

The general utility or purpose of the methods is concisely covered. Such information has been lacking in the official treatises of the validating associations.

Sections are present that cover the activities of the technical associations in methods development and validation, pertinent regulatory legislation and regulations, methods pertinent to nutrition labeling, the importance of methods to specifications and standards, the Food Chemical Codex, and the significance of precision and accuracy.

ACKNOWLEDGMENTS

Appreciation is expressed to Dr. Bryan L. Madison of the Procter & Gamble Company in Cincinnati, OH, Dr. David Firestone, FDA in Washington, and Nancy Palmer, AOAC in Arlington, VA.

References

1. Firestone, D. et al. 1991. Food Technol. 45(2), 90.
2. Bigelow, S. W. 1991. Food Technol. 45(5), 88.

3. Baur, F. J. and Parker, W. A. 1984. J. Assoc. Official Anal. Chemists *67*(1), 3.

4. Steiner, J. 1993. Inform *4*(3), 308.

5. Steiner, J. 1993. Inform *4*(6), 706.

6. Madison, B. and Roberts, B. 1991. Inform *2*(8), 729.

7. Berner, D. L. 1992. Inform *3*(7), 824.

8. White, P. J. 1991. Food Technol. *45*(2), 75.

9. Smith, L. M. et al. 1986. J. Am. Oil Chemists Soc. *63*(8), 1017.

10. Steiner, J. 1993. Inform *4*(2), 211.

11. *Health Aspects of Dietary Trans Fatty Acids*, Life Sciences Research Office, Federation of American Societies for Experimental Biology, August (1985) Bethesda, MD.

12. Ratnayake W.M.N. et al. 1990. J. Am Oil Chemists Soc. *67*(11), 804.

13. Firestone, D. 26, October 1992. Private communication; see also Firestone, D. 1993. Inform *4*(2), 211.

14. W. W. Christee, 1992. Advances in Lipid Methology-one The Oily Press, Ayr, Scotland 273–322. 1993.

15. Steiner, J. 1993. Inform *4*(4), 455.

16. Mossoba, M. M. 1993. Inform *4*(7), 854.

17. Nollet, L. M. 1992. *Analysis by HPLC*. New York: Marcel Dekker; DeMan, J. M. 1991. *Encyclopedia of Food Science and Technology, Vol. 2*, Y. H. Hui (Ed.). New York: John Wiley and Sons.

18. DeMan, J. M., Aurand, L. W., Woods, A. E., and Wells, M. R. 1987. *Food Composition and Analysis*. New York: Van Nostrand Reinhold.

19. Linscheer, W. G. and Vergroesen, A. J. 1988. *Lipids in Modern Nutrition and Health and Disease*, M. E. Shils and V. R. Young (Eds.). Philadelphia: Lea and Febiger.

20. National Academy of Sciences. 1990. *Nutritional Labeling: Issues and Directions for the 1990s, Report of the Committee on the Nutritional Components of Food Labeling, Food and Nutrition Board*. National Academy Press, Washington, DC:

21. Caccomandi, V. 1990. Official J. Eur. Community *276*, 40.

22. United Kingdom Department of Health. 1991. *Dietary Reference Values for Food Energy and Nutrients for the United Kingdom, Report of the Panel on Dietary Reference Values of the Committee on Medical Aspects of Food Policy*. London: Her Majesty Stationery Office.

23. The Referee. 1993. *17*(2), 6.

24. The Referee. 1993. *17*(5), 1.

25. De Vrios, J.W. and Nelson, A.L. 1994. *Food Technol.* 48(7), 73.

26. Caro, R. H. and Morgan, W. E. 1991. Food Technol. *45*(7), 62.

27. Belbin, A. A. 1993. Inform *4*(6), 648.

28. Giese, J. 1993. Food Technol. *47*(5), 88.

29. Cole, C. F. and Urthoefer, F. T. 1993. Inform *4*(4), 432.

General References

Applewaite, T. W. Ed. *Bailey's Industrial Oil and Fat Products, Vol. 3*, New York, Wiley-Interscience, (1985).

Swern, D. Ed. *Bailey's Industrial Oil and Fat Products, Vol. 2*, New York: Wiley-Interscience, 1982.

AOCS Official Methods and Recommended Practices, 4th ed., 1989 and subsequent annual supplements.

AOAC Official Methods of Analyses, 15th ed., 1990.

Perkins, E. G. (Ed.). 1991. *Analyses of Fats, Oils and Lipoproteins*. Champaign, IL: AOCS.

Rossell, J. B. and Pritchard, J. L. R. 1991. *Analysis of Oil Fats and Fatty Foods*. New York: Elsevier Science Publishing Co.

Hamilton, R. J. and Rossell, J. B. 1986. *Analyses of Oils and Fats*. New York: Elsevier Applied Science Publishers.

Hart, F. L. and Fisher, H. 1971. *Modern Food Analyses*, New York: Springer-Verlag.

Williams, K. A. 1966. *Oils, Fats and Fatty Foods: Their Practical Examination*, New York: American Elsevier Publishing Co.

Chapter 15

Food Product Development

This chapter could also be entitled. "Industrial Food Research and Development." Activities in this area can include the handling of problems and complaints (reactive firefighting), quality assurance programs including quality function deployment (QFD or TQ) and total quality management (TQM), proactive studies to prevent problems, process and engineering development, products research, and systematic approaches to the development of new products and to the improvement of current products.

Improvements of current products may include package improvements, ingredient changes, processing changes, flavor improvements, and other changes that improve manufacturing and distribution efficiency or offer more economic advantages to the user. A good solid Research and Development (R&D) program is essential to the growth and future health of all food processors and others involved with the use of oils and fats. In the foodservice field this activity is also important. As an example, fast-food chains are continually trying to develop new sandwiches, new entree items, and, especially today, items of improved nutritional characteristics that will give a competitive advantage and an increase in market share. In addition, they must conduct ongoing studies of both current customers' preferences and future or potential customers' preferences.

For the smaller food processor or individual independent foodservice operator, a large new product development program may not be practical, but the search for new products, quantity food recipes, improved or more efficient methods of preparation, and improved methods of serving are of utmost importance to the future of the business.

In this chapter, some methods in use by large food processors and chain foodservice operators will be presented. Smaller operators should be able to find some ideas that can be adapted to their operations on a smaller scale. It should be understood that there are also other ways of approaching product development and recipe development activities. In addition, many small food processors and even some large foodservice chains contract out portions or all of their R&D programs with companies that have long-established R&D facilities. A knowledge of these programs and how they work (or should work) will be helpful in monitoring the progress and success of such contracted programs.

THE DESIGN OF SUCCESSFUL NEW FOOD PRODUCTS

Successful products are designed and developed to meet the particular needs and taste of a large number of customers over some span of time. The method of introducing a new product to the market includes a thorough pattern of development, a market testing period, and expansion to a national business. This method of market development succeeds best when there is an already existing customer demand that can be stimulated by a new product. No matter how appealing the product, if an initial demand does not exist, it will probably be unreasonably costly to keep it in the market long enough to develop whatever latent strength of demand there might be.

Objectives of a Good Product Design

The requirements for a successful product are as follows:

1. The outstanding performance of a needed and desired function for a large number of customers. For example, many recent consumer surveys have indicated that flavor/taste of food products is the most important attribute.

2. The satisfactory performance of any secondary functions and the absence of important negative features. An example could be a nutritional product advantage that would not create any serious negative features.

3. Properly balanced esthetic qualities that will appeal strongly to the tastes of a large group of customers. These could include a good balance between flavor/taste, color, and aroma.

4. At least one distinctive feature to set it apart from competing products and give a strong promotional message. For example, product convenience is a very important feature for single persons and the two-wage-earner families.

5. A cost that will return a normal profit when the product is priced in proper relation to quality advantages that are easily recognized by a significant number of customers/potential customers. A new product must, of course, be profitable, and when it has a price advantage (readily recognized value for the price), it should be a winner.

In addition to the above, nutritional plusses must be considered for many products.

Procedures for Designing a New Product

Positive Function

First establish and then satisfy the essential functional requirements. There are often both primary and secondary functions. Reproduce under controlled conditions as many as possible of the physical situations in which the product will be used. By direct observation, see what the users will see and taste what the users will taste. This requires ingenious development of circumstances that are typical of those the customers encounter when using the product.

In addition to direct observation, market research is a prime source of information about customer practices. This information must be used with caution because what people say they do or like may not actually describe what they truly do or like. This is especially true when customers are asked to rate products that are provided free for testing.

A word of particular caution applies to the development of food products. Market research must take into account demographic preferences as well as the difference in likes and dislikes of men, women, and children. Children often have tastes that are different from those of their parents. In addition, convenience may be appealing to the foodservice owner or manager but may have little or no value to foodservice customers. If the eating quality is not superior, convenience alone represents a poor basis for a new product.

Elimination of Negative Functional Defects

Among other secondary characteristics, it is, of course, necessary to satisfy any legal and toxicological requirements. Professional help may be necessary for getting answers to questions on such requirements. So far as unexpected and peculiar negatives are concerned, it is often fruitful to set up in one's imagination the difficulties most likely to be experienced with the product and then to search for circumstances in which these difficulties might develop. In the case of a product sold to thousands of customers over a long span of time, if trouble can be experienced, it probably will be.

It should be remembered that the package represents the first exposure and the first impression of the product. It also affects the way a product is handled and used. The work on product and package proceeds independently, but possible interactions between the two are among the secondary performance characteristics that must not be overlooked.

Customer "Appeal"

We come finally to the matter of esthetic characteristics. The reactions to odors and flavors are instinctive and largely below the level of conscious attention. They can influence the total reaction without being consciously registered. Reactions to color are more conscious but also involve large elements of instinct and intuition, especially when food colors are involved. In addition, with such products as peanut butter, cakes, pies, and fried foods, the texture and mouth sensation are vitally important.

There are divergencies in taste/flavor. See Chapter 16 for a more complete discussion. In some products such as shortenings and oils, we are generally aiming at the total absence of any specific flavor. This is simply measured by direct observation. In other products, however, we intend to use a particular flavor as the principal point of difference or preference. We are trying to make a strong appeal to some portion of potential customers. Such characteristics must be appraised by comparisons with other products that have had a history of acceptance. Panel and survey techniques form the basis for these conclusions.

Keep in mind that novelty tends to be more appealing in an initial exposure, such as in a blind test, than it is in the actual market situation. For example, tests in which a yellow color is used rather than white in shortening demonstrate a consistent consumer preference for yellow, but white shortenings still dominate the market. On the other hand, yellow color is a must for butter and margarine. A novel aesthetic feature may be used to dramatize more basic differences but rarely will it support a product without there being a more substantial reason for existence.

A feature strongly liked by some may be strongly rejected by others. There are differences in individual physiology and in individual taste. There are, in fact, certain chemicals to which the sensitivity of individuals varies widely. Negative reactions, in particular, are often difficult to reduce to quantitative terms, and we often must rely on clues that can be obtained from panel tests and market research. Even if only one or two individuals in a group of 200 or 300 express strong dislike for some particular quality, it should be a matter for serious attention; its elimination may not be possible, but its existence is a fact to be noted.

It is very important to communicate with all interested parties for input at regular intervals throughout the product development stage. This can include representatives from top management, packaging, process development, manufacturing, advertising/promotions, sales, buying, and those who represent the organization with the various regulatory agencies.

Final Evaluation

A good product design is the result of a combination of technical and esthetic qualities that is often worked out experimentally by one person. This design must have "integrity"; that is, it must fit together in a way that will satisfy the needs and tastes of a large number of people. A designer should never release any product for market research appraisal without having a strong conviction, based on personal use, that it has a right to succeed. Market research is a useful but not a completely adequate tool for the final appraisal of a product design. It is not a substitute for an inner conviction based on intimate knowledge, personal experience of repeated exposure, and many conversations with others who have used the product experimentally.

The interpretation of data acquired by market research is an art in itself. No single test stands alone. It should be interpreted within the total context, which includes all other knowledge about the product and about the test techniques that have been used. The viewpoints of market research personnel and advertising personnel are valuable sources of insight.

The appearance of any negative in a customer test is reason to pause and review. The failure of a negative to appear in a test does not necessarily exclude it as a factor in the market. For this reason, any test market needs to be carefully followed and each complaint needs to be understood. It may be necessary to seek a follow-up interview with the complainant. Complaints should be followed up until they are understood.

Aesthetic factors pose complex problems. In the usual two-product "blind test" we look for information in the "reasons for preference." The interviewing technique forces users to verbalize as "reasons" impressions that are not ordinarily consciously registered. This leads to a certain distortion. Immediately obvious differences such as color or physical form tend to be reported with some exaggeration of their ultimate significance, whereas differences of flavor and texture, which are more subjective in nature, may be undervalued. Often these reactions are translated into "apparent" functional differences. It is necessary to compensate for such distortion by comparing the results of a single test with those of other tests in which similar differences were studied. There is no formula for doing this. Each product has its own pattern.

In developing a product that is competing directly with a product of known market performance, a two-product blind test is often the best measure available of the first impression the product makes on the present user of the competing product. A good impression in the initial trial will, in large measure, determine the likelihood that there will be a second trial. It is necessary to make this good first impression, but this in itself is not enough—the important fundamental properties must have been designed into the product from the beginning. Finally, there is no substitute for market testing, but this is so expensive that it is necessary to be sure in one's own mind that all other channels of evaluation have been exhausted and that the product has a good design before any marketing is undertaken.

There are many aids to judgment, but there is no substitute for it. Experienced judgment is required at this point.

OTHER FACTORS IN NEW PRODUCT DEVELOPMENT

The new product development manager's job can be fascinating. It can also be discouraging and is often very risky. A good new product manager is alert to customer changes in wants and may have to constantly battle short-range company objectives. This manager must be flexible and objective, must never give up on the right idea or right product, and must be a self-starter.

To be successful in new product work, the manager must be thoroughly committed to the project, must understand company policy and objectives, and must know how to sell the right programs and how to obtain top management support.

The manager must be able to define both short-range product development goals and long-term objectives. With some companies, product development/research and development play an integral and dominant role, and this is expected to succeed. Even in a seemingly ideal corporate atmosphere, the new products manager is still charged with the responsibility of selecting those activities that have the best chance of success; top management still must be sold on which projects to finance. Whenever money must be allocated well before returns and profits can be realized, getting that money is bound to be a difficult job even in the best of corporate climates.

Sometimes new products and new opportunities come through acquisitions rather than through internal generation. Top managers often prefer the acquisition route because some modest returns may come almost instantly and they can better visualize routes to improving current products, expanding the line, and providing more effective marketing programs. The acquisition route can alleviate the normal top management impatience and need to obtain quicker results. In today's world it becomes increasingly more important to move faster. The timeline for a basic technical understanding is shortened, and new products can move into the marketplace at a faster pace.

The new product manager does have an advantage over top management if the product development job has been done right. If the idea and the product are right, the cost information is reliable, the legal and health considerations have been correctly addressed, and pretesting and customer market testing have been done correctly, the manager has all of the data needed for communicating with top management.

SUMMARY

Successful operations, large and small, must be constantly alert in order to keep on being successful. Therefore, all must have a program to assure continued improvement of products and services to better meet their customers' needs.

References

1. Sloan, A. 1994. Food Technol. *48*(1), 36–37.

Chapter 16

Flavor

Fred J. Baur, Ph.D.

INTRODUCTION

The chapter on nutrition (Chapter 13) contains a section on the various desirable contributions oils and fats make to formulated foods. With oils and fats, most uses require bland products. Oil processing is designed toward removal of flavor and prevention of flavor formation. There are exceptions, of course. The natural flavors of olive and peanut oils are desired by some consumers for salad and cooking purposes. In addition, flavored oils and shortenings, as with added butter flavor, are commercially successful.

Oils and fats do, however, have important roles in the flavor or taste of foods. They influence texture and mouth feel and they serve as a carrier and heat-transfer medium for flavors. Do understand that the presence of oil or fat delays the volatility of aromatic substances so important to flavor as well as coating the taste receptors in the mouth, thereby delaying responses to stimuli.

Whereas component oils or fats may not influence finished product selection via the broad contribution they make to eating quality, they can and do contribute to rejection via disagreeable and unacceptable flavors. (See chapter 6.)

WHAT IS FLAVOR?

Flavor is the key quality feature determining the continued acceptance of a food by the consumers.

Webster's Collegiate Dictionary's definition of flavor includes (1) odor or fragrance, (2) the quality of anything that affects the taste, and (3) a substance that flavors and includes taste as a synonym. The definition of taste speaks to touching, as by touching with the tongue.

Much more pertinent to this text is the definition by the U.S. Society of Flavor Chemists which says flavor is the sensation caused by those properties of any substance taken into the mouth that stimulates one or both the senses of taste and smell, or the general pain, tactile, and temperature receptors in the mouth (1).

Taste has been defined as the combined sensory responses of smell (olfactory), flavor (gustatory or taste), and tactile (mouth feel or touch). In fact, both consumers and marketing personnel are increasingly using the word taste (2). So, be aware that the words flavor and taste can be used interchangeably. Keep in mind also that added flavors are primarily made of volatile materials; hence, the buccal cavity receptors may not come into play.

In past usage, flavor is a combination of aroma and taste, and taste is comprised of the tongue's responses to salty, sweet, sour, and bitter sensations. (The literature has included a metallic sensation. This has not been commonly accepted). But, the tongue surface also reacts to tactile and temperature stimuli. Prime are texture/viscosity, astringency (alum, tannins), cooling (menthol), heat or bite (red pepper or piperene), and burn (horseradish). All contribute to the overall flavor perception, with the most important or dominant characteristic being aroma. The flavors perceived in oils and fats, such as beany, oxidized, rancid, buttery, fishy, and so on, are largely detected by smell not by the taste buds.

The oral or buccal cavity, primarily the tongue, contains some 9000 taste buds. Those that pick up salty and sweet sensations are located at the tip and edges of the tongue, sour at the rear edges, and bitter at the back of the tongue.

Odor or aroma is assuming an even greater role in flavor. This results from increased sophistication of product base through the use of components such as highly functional emulsifiers and carbohydrate fat replacers. These tend to tie up volatile flavor and odorant materials. The aroma attribute can be another indicator of difference in flavor character or profile mainly due to the release of volatile flavorants, which can mimic the release of these same flavorants during mastication.

HISTORY OF ADDED FLAVORS (3)

A bit of history on the use of flavoring materials may be of interest.

The origin of their use is lost in antiquity. The probable first use was that of spices in rice in China. The use of cassia for that purpose is described as early as 2700 BC. There are biblical references to the use of perfumes and spices; spice merchants bought Joseph from his brothers about 1730 BC. The Egyptians used spices in their foods. The Greeks and Romans also used extracted oils.

In medieval times, spices were used for food preservation and techniques developed for the distillation of essential oils.

Chocolate and vanilla were brought back to Europe by Cortez in about 1520.

In the middle to late 1800s the potential of synthetic aromatics to increase the effectiveness of natural flavors was beginning to be realized. In 1851 at the World's Fair in London, solutions of esters were exhibited and recommended as artificial fruit essences. By 1875 the United States was importing imitation fruit rum and brandy flavors from Germany. The growth of the flavor industry continues to increase 3–4% per year.

THE IMPORTANCE OF FLAVOR

It has been mentioned that the flavor of a product is considered the prime determinant in its acceptance. It also needs to be realized that poor or off flavor will cause product failure. The major cause of consumer complaints to the food industry is off flavor, and products are withdrawn annually from the market because of this defect.

Not surprisingly, the level of flavor as well as the flavor itself is important. Children and older people, and the percentage of the latter continues to increase, tend to prefer milder foods.

Flavors will continue to play an important role in increasing the available food supplies by making acceptable foods from unusual sources as algae and biotechnology.

FLAVORS OF OIL AND FAT PRODUCTS

The flavors of oil and fat products are generally ascribed to the component fatty acids, to the unsaturated fatty acids, and to the esters of the acids, aldehydes, and related compounds. More specifically, the flavor-related volatiles mostly come primarily from degradation of hydroperoxides formed by oxidation of the unsaturated fatty acids.

Oxidation can be auto-oxidation, photo-oxidation, and/or enzymatic oxidation.

Auto-oxidation proceeds through free radical chain reactions via the attack on the alpha methylene to the double bond.

Photo-oxidation is a much faster reaction that catalyzes the attack at the double bond by singlet oxygen formation.

Enzymatic oxidation of lipids is caused primarily by the lipoxygenases. Plant lipoxygenases produce hydroperoxides that produce the secondary products important to flavor of many food products besides oils and fats. As examples, mushroom flavor is believed to be mainly 1-octen-3-ol. Tomato flavor's primary components are *cis*-3-hexenol, *cis*-3-hexenal, and *trans*-2-hexenol. All of these substances come from linoleic acid, as does 2,4 decadienal. This is a breakdown product of oxidized linoleic acid and is related to the desirable fried food flavor typical of potato chips.

The compounds identified in the volatiles of oxidized soybean oil include aldehydes, hydrocarbons, both aliphatic and aromatic, ethers, ketones, lactones, and so on.

MEASUREMENT OF FLAVORS OF OIL AND FAT PRODUCTS

Measurement of flavors can be by sensory or organoleptic means or by chemical and instrumental means.

Insofar as product acceptance is concerned sensory is the method of choice. The reasons are primarily that (1) the sense of smell is more sensitive than instrumentation and (2) the sensory acceptance or rejection, or rating of a flavor, represents the real world in that it relates to what the consumer wants.

Chemical/instrumental means have value in that they may forewarn of possible instability or poor processing.

The primary disadvantage of sensory testing is that it is subjective, and considerable training may be required to yield acceptable accuracy and precision.

OBJECTIVE METHODS FOR ASSESSING OIL QUALITY (1)

The most prevalent cause of poor flavor in oils and fats is oxidation, which leads to aldehydes, ketones, alcohols and other compounds that affect flavor. Methods were, therefore, designed to measure the degree of oxidation or certain byproducts.

Peroxide Value (PV): a measure of the primary products of oil oxidation, the peroxides. The peroxides themselves have no flavor and break down readily. Correlation of measurement and flavor is variable.

Anisidine Value: determines level of aldehydes, primarily 2-alkenals that are derived from peroxides. The Totox value equals the anisidine value plus twice the PV. This sum is felt to be a measure of past and current oxidative "history."

Thio-barbituric Acid Test: measures malonaldehyde, which has been shown to correlate well with PV of oils containing fatty acids of three or more double bonds.

Conjugated Diene and Triene: double bonds conjugate as oxidation occurs; of value more with oils of known composition.

volatiles: direct measurement by gas-liquid chromatography of responsible compounds. This approach continues to receive considerable attention.

Predictive Tests (1): oil quality predictive tests that have been applied include the Active Oxygen Method (AOM), ASTM (American Society for Testing and Materials) Oxygen Bomb, Differential Scanning Calorimetry (DSC), and Thermogravimetric Analyses (TGA).

Unfortunately none of the above has shown good correlation with oil flavors.

SENSORY TESTING

Criteria for taste panels are as follows:

1. Taste panels should be conducted in a well-ventilated, lighted, and air-conditioned room.

2. Each panel member should have a separate booth where, in a comfortable sitting position, he or she has easy access to the samples. Experience has shown that quietness, orderliness, and regularity contribute to accurate evaluations.

3. Odors and flavors of oils are more easily detected if the oils are warm; 43–49°C (110–120°F), with 44°C (112°F) recommended for the actual tasting. This temperature should be carefully controlled from sample to sample.

4. Samples should be presented "blind" to the panel, in pairs, for most accurate results, but many panels can handle sets of four at one sitting.

5. Water at body temperature or slightly higher should be used for rinsing the mouth between samples. No sample should be swallowed.

6. Oils with the strongest flavor should be tasted last; this would normally be determined by odor. Some statistical plans require that samples be tasted as presented.

7. Should an obnoxious flavor be encountered, the panelist should wait at least 10 min before examining the next sample.

8. Heavy smokers should be avoided in panel selection. Testing by nonheavy smokers should be delayed at least 1 h after indulgence.

9. Nasal congestion is a reason not to use an individual, depending on the severity.

10. At least 1 h should be allowed to lapse after a meal, consumption of coffee, soft drinks, and other flavored beverages, and the use of chewing gum, prior to testing.

11. An expert panel should consist of at least three to five qualified panelists.

12. Consumer-type panels for description or difference panels may run as high as 20 members.

13. No more than four samples should be flavored in a paneling session. If a larger number of samples must be tested, allow 30 min recovery time for each group of four samples.

14. Human relations are an important part of any taste panel. Rewards at the end of the taste period, such as cookies or cake, are helpful. Where possible, sharing of research results can help sustain the interest of the panel members.

If at all possible, the samples should be smelled before tasting and then tasted in order of increasing perceived odor. Smelling is done by inhaling through the nose for a period of 2–3 s with both nostrils open. Sniffing should be cautious. Use the same nostril or both for direct comparisons because the airstream is rarely equal between the nostrils. Do not use the hands to hold the samples as skin odors are usually present. Be aware that the sense of smell recovers much more rapidly than does taste.

In tasting, slurp the oil to help aerate the sample to waft the aroma into the nose. The sample should be retained in the mouth long enough for the panelist to make an accurate judgment. This period will vary with the individual and the sample.

Oil flavor descriptions for the most common flavor types are as follows:

Bland: having little, if any, flavor or odor. The flavor is mild and neither irritating nor stimulating.

Nutty: flavor reminiscent of fresh pecans—one of the least objectionable flavors. When very intense, it appears as an objectionable, rubberlike flavor.

Buttery: suggests the aroma and flavor of fresh, sweet butter. A strong buttery flavor would refer to a pronounced, sweet butter flavor, not old or rancid.

Beany: a flavor characteristic of soybean oil products—somewhat objectionable in slight degree. The intense characteristic is termed "weedy."

Oxidized: a flavor characteristic of oils exposed to air, particularly cottonseed oil. Sometimes appears to be "metallic" to some observers. A strong oxidized flavor is termed "rancid."

Raw: a flavor characteristic of underdeodorized oils that is frequently observed in cottonseed and soybean salad oils; more intense characteristic is sometimes termed "musty" or "earthy."

Grassy: the "green" flavor suggesting the astringency or bitterness of green grass. When samples of oil have been exposed to daylight, the predominating flavor is usually described as grassy. If the flavor is strong, some tasters modify this term by describing it as peppery or biting. Others consistently describe this flavor, which the majority call grassy, as metallic.

Reverted: a flavor characteristic of slightly old soybean oil or of one exposed to air. A more intense characteristic is termed "painty" and/or "fishy."

Rancid: a very disagreeable, sometimes sharp, biting, or nauseating flavor in very old or strongly oxidized fats. It is related to "oxidized"; very objectionable in any degree, and nauseating when strong.

Painty: a flavor reminiscent of the odor of linseed oil, or drying paint, after the solvent odor has disappeared. It is related to "reverted"; quite objectionable in slight intensity to very objectionable in strong intensity; can be nauseating when very intense.

Watermelon: flavor reminiscent of biting into watermelon rind, or of cucumbers; sometimes seen in overaged, hermetically sealed cans of shortening. The flavor is quite objectionable when intensity is slight to very objectionable when in strong intensity. It is also related to "reverted."

Fishy: a flavor reminiscent of cod-liver oil, sometimes seen in heat-mistreated soybean oil. It is related to "reverted"; quite objectionable in strong intensity and can be nauseating when very intense.

There are many sensory test methods. The fundamental types are preference-acceptance, discriminatory, and descriptive.

Preference tests normally are the paired comparison, the hedonic scale, and ranking. The single product paired comparison test is run by presenting samples individually and removing after each evaluation so there is no opportunity for side-by-side comparison. This is a more realistic and often more sensitive method of two product comparison.

Discriminatory or difference tests include the triangle test, simple paired comparison test, duo–trio test, and multiple comparison test. The duo–trio and triangle are two of the most proven and reliable. The triangle test consists of three coded samples, two identical and one different. The panelist is asked to identify the odd sample. In the duo–trio, three samples are presented with one being a known coded reference. One of the other two samples is identical to the reference and the other different. The panelist is asked to identify the different sample. This test requires less tasting but is not as efficient. The multiple comparison test consists of presenting a known reference sample along with several coded ones. The panelist is asked to compare the coded samples with the known for some specific characteristic.

The descriptive tests are the most complex and require the most training. Examples include flavor profile, texture profile, and attribute grading. In the last, flavor attributes known to exist in the product, such as buttery, rancid, and so on, are listed and the panelist is asked to give a numerical score according to the intensity found. The test is good for measuring degrees of difference in known attributes. It requires a panel trained specifically for the project and training can take up to 4 weeks.

A typical sensory rating sheet, for french fried potatoes, is shown in Fig. 16.1.

Panel Supr. _____ Test No. _____ Class of Food _____			Sensory Testing French Fried Potatoes		
Color	Very light Dull/	Light	Good color	Dark Slightly	Very dark Bright/
Appearance	unappetizing	Slightly dull	So-so/ordinary	burnt	appetizing
Texture	Brittle/dry		Crisp and most		Limp/soggy
Browned flavor	Raw/underdone	Slightly underdone	Well done	Slightly overdone	Burnt
Greasy taste	None	Slight	Moderate	Strong	Excessive
French fry flavor	Poor	Fair	Good	Very good	Full/natural
Off flavor notes	None	Slight	Moderate	Strong	Extreme
Overall rating	Poor	Fair	Good	Very good	Excellent

Descriptive Flavor Terms Noted

Earthy	___	Stale	___	Bitter	___	Grainy	___	
Raw	___	Starchy	___	Oxidized	___	Gummy	___	Nutty ___
Cereal	___	Musty	___	Beany	___	Grassy	___	Tallow ___
Potato skins	___	Cardboard	___	Buttery	___	Lardy	___	Other ___

FIGURE 16.1.   Sensory testing sheet: french fried potatoes.

A recent *Food Technology* article (June 1993) reviews the main sensory test methods, panel selection, and statistical designs (4).

FLAVOR CHANGES ON DEEP FAT FRYING

The mechanism of flavor development in heated oils is essentially that of lipid oxidation. As was mentioned earlier, this leads to the formation of the peroxide radical, then the formation of hydroperoxide, which decomposes to form volatile flavor compounds. The products of thermally, induced oxidations may have differences from the typical lipid oxidation products formed at ambient temperatures. Not surprisingly, at elevated temperatures more sites for oxidation become available, which means a wider range of end products may occur. The last of volatiles identified is composed mainly of acids, alcohols, aldehydes, hydrocarbons, ketones, lactones, esters, aromatics, and miscellany like pentyl furan and 1,4,dioxane. Two hundred twenty volatiles have been identified in deep fat frying.

The water in a food being fried will lead to hydrolysis of the oil, forming fatty acids which may add to flavor, steam distillation, with a partial loss of the volatile flavors present, and depending on the composition of the food, the creation of other flavors.

OFF FLAVORS

Off flavors within foods mainly result from lipid oxidation, nonenzymatic browning, enzymatic changes, and photocatalyzed reactions.

Lipid oxidation is a continuation of the processes taking place during storage, exposure, and/or heating. Similar chemical compounds are formed.

Lipases and lipoxygenases are naturally present in foods and are not always inactivated by processing or heating. Lipoxygenases lead to oxidation and the lipases cause hydrolyses of the triglycerides to produce free fatty acids.

Oxidized notes can be oxidized, rancid, old, bitter, metallic, and so on. Hydrolytic notes can be soapy, bitter, fishy, green, and so on.

Decadecenal and some other fatty aldehydes can be perceived odorwise very early in the onset of rancidity.

SUMMARY

The definitions of flavor and taste are described. The role and importance of oils and fats in the flavor of foods as well as their own innate flavors are discussed. Fresh, exposed, and oils used in deep fat frying are considered along with the creation of various flavor artifacts.

Mention is made of the important role testing, both sensory and chemical/instrumental, plays in flavor assessments. Examples are given.

References

1. Applewaite, T.W. 1985. *Bailey's Industrial Oil and Fat Products, Vol. 3*, New York: Wiley-Interscience.

2. Sloan, A.E. 1993. Food Technology 47(6), 46.

3. Apt, C.M. 1978. Flavor: Its Chemical Behavioral and Commercial Aspects. Westview Press, Boulder; CO:

4. Lawless, H.T. and Claassen, M.K. 1993. Food Technol. 47(6) 139.

General References

Min, D.B. and Smouse, T.H. 1985. *Flavor Chemistry of Fats and Oils*. American Oil Chemists Society, Champaign, IL:

Heath, H.B. and Reineccius, G. 1986. *Flavor Chemistry and Technology*. AVI–Van Nostrand Reinhold Co. New York:

Warner, K. 1991. *Assesment of Fat and Oil: Quality: Sensory Methodology, Analyses of Fats, Oils and Lipoproteins*, E.G. Perkins (Ed). American Oil Chemists Society, Champaign, IL.

Index